Recent Advances in Renewable Energy
(Volume 4)

Solar Chimney Power Plants: Numerical Investigations and Experimental Validation

Authored by

Haythem Nasraoui, Moubarek Bsisa,
&
Zied Driss

Laboratory of Electromechanical Systems (LASEM),
National School of Engineers of Sfax (ENIS), University of Sfax (US),
B.P. 1173, Road Soukra km 3.5, 3038, Sfax, Tunisia

Recent Advances in Renewable Energy

Volume # 4

Solar Chimney Power Plants: Numerical Investigations and Experimental Validation

Authors: Haythem Nasraoui, Moubarek Bsisa & Zied Driss

ISSN (Online): 2543-2397

ISSN (Print): 2543-2389

ISBN (Online): 978-981-14-6175-0

ISBN (Print): 978-981-14-6173-6

ISBN (Paperback): 978-981-14-6174-3

Published by Bentham Science Publishers Pte. Ltd. Singapore. All Rights Reserved.

need for a court order if at any point you breach any terms of this License Agreement. In no event will any delay or failure by Bentham Science Publishers in enforcing your compliance with this License Agreement constitute a waiver of any of its rights.

3. You acknowledge that you have read this License Agreement, and agree to be bound by its terms and conditions. To the extent that any other terms and conditions presented on any website of Bentham Science Publishers conflict with, or are inconsistent with, the terms and conditions set out in this License Agreement, you acknowledge that the terms and conditions set out in this License Agreement shall prevail.

Bentham Science Publishers Pte. Ltd.
80 Robinson Road #02-00
Singapore 068898
Singapore
Email: subscriptions@benthamscience.net

**BENTHAM
SCIENCE**

CONTENTS

PREFACE

This book aims to study the effect of the geometrical parameters on the airflow behavior inside a solar chimney power plant.

In the first part, we have developed simulation by using the CFD software ANSYS Fluent to model the airflow. In these conditions, we have adopted the realizable k-ε turbulence model, the DO radiation model, and the convection heat flux transfer model. These models have been validated with anterior experimental results due to the acceptable coherence between results. In the second part, alternate geometric configurations of the solar chimney power plant were numerically studied to expand on the design optimizing of the solar chimney. The goal is the study of the effect of the geometric parameters on the airflow behavior inside the solar chimney to obtain an optimal size available to construct a prototype of a solar chimney power plant. The developed study confirms that the increase in the height and diameter of the chimney, and the diameter of the collector increases the temperature and the air velocity. However, an increase in the collector height decreases these parameters. An experimental study is also presented in the last part of this book. The experimental prototype, constructed at ENIS, is used to study the environmental temperature, distribution of the temperature, air velocity, and the power output generated by the turbine. The main results were found from this prototype are the solar radiation and the gap of temperature in the collector. These parameters are important factors affecting the performance of the solar chimney power plant.

ACKNOWLEDGEMENT

The authors sincerely and heartily acknowledge their colleagues from the Laboratory of Electro Mechanic Systems, for the assistance and advice relating to this book. Particularly, they would like to express their sincere gratitude to Dr. Ahmed Ayadi, Dr. Abdallah Bouabidi, and Prof. Mohamed Salah Abid for their continuous help and support.

CONSENT FOR PUBLICATION

Not applicable.

CONFLICT OF INTEREST

The author(s) confirm that this chapter contents have no conflict of interest.

Haythem Nasraoui, Moubarek Bsisa & Zied Driss
Laboratory of Electromechanical Systems (LASEM)
National School of Engineers of Sfax (ENIS)
University of Sfax (US),
B.P. 1173, Road Soukra km 3.5, 3038, Sfax
Tunisia

Nomenclature

g Gravitational acceleration (m.s^2)

H Chimney height (m)

d Chimney diameter (m)

D Collector diameter (m)

h Collector diameter (m)

θ Collector slope (°)

r, z Cylindrical coordinates

V Velocity (m.s^{-1})

V$_c$ Air velocity at collector chimney inlet (m.s^{-1})

p Static pressure (Pa)

p$_d$ Dynamic pressure (Pa)

p$_i$ Pressure inlet (Pa)

p$_o$ Pressure outlet (Pa)

h overall heat transfer coefficient (W.m^{-2}.K^{-1})

T Temperature (K)

T$_0$ Ambient temperature (K)

ΔT Air temperature between the collector inflow and outflow (K)

A$_{ch}$ Chimney cross-section area (m^2)

A$_c$ Collector area (m^2)

C$_p$ Specific heat capacity of the air (J.kg^{-1})

m˙ Mass flow of air (kg.s^{-1})

ρ Density of the air (Kg.m^{-3})

ρ$_c$ Specific density of air at the chimney inlet (kg.m^{-3})

Q$_s$ Total energy input in the collector (W)

Q Energy produced by the collector (W)

η$_c$ Collector efficiency

η$_{ch}$ Chimney efficiency

η$_t$ Turbine efficiency

η$_g$ Generator efficiency

η$_{tg}$ Efficiency of the turbine generator

η Overall efficiency of the solar chimney

G Global radiation (W.m^{-2})

  α Absorptivity of the collector roof

  τ Transmissivity of the collector roof

ΔT$_a$ The average difference of temperature under the collector and the ambiant (K)

  U Voltage (V)

  I Current (A)

  P$_e$ Electrical power (W)

  k Turbulent kinetic energy (m^2.s^{-2})

  ε Dissipation rate of turbulent kinetic energy (m2.s^{-3})

  μ$_t$ Turbulent viscosity (Pa.s^{-1})

  μ Dynamic viscosity (m^2.s^{-1})

Cμ Function of mean flow and turbulence properties

$\tilde{\Omega}_{ij}$ The mean rate-of-rotation tensor viewed in a rotation reference frame

  τ$_{ij}$ Molecular stress tensor for Newtonian fluids

  Γ$_\phi$ Diffusion coefficient for the scalar quantity ϕ

  S$_\phi$ Source term for the ϕ equation

Introduction

Nowadays several energy sources are utilized on a large scale around the world such as oil, gas, and nuclear. Since the oil crisis, depletion of fossil fuel reserves, global warming, and other environmental concerns and continuing fuel price rise. For these reasons, the existing sources of conventional energy may not be adequate to meet the ever-increasing energy demands. Moreover, the demands for energy will tend to grow and it draws attention to the need to save energy, including through the use of soft power. Consequently, Engineering was responsible for finding new and different energy sources to fossil fuels to move them from the production of industrial energy and one day replace. The features that are mainly looking for a source of energy are to produce the least environmental impact possible and less pollution of any kind, which has a potential of energy efficiency and acceptable, that is safe, simple, reliable, cost-effective and cheap, that is renewable energy.

There are many forms of renewable energy resources that are currently available for integration into the power grid; the top four energy sources are wind, sun, water, and geothermal. For thousands of years, civilizations have been harnessing the Sun's energy, so there are several ways of solar systems. One of these ways for humans to harness the sun's light for energy production is called solar chimney technology, also called solar towers, to avoid confusion with polluting industrial chimneys. In this context, we are interested in developing this technology.

This book is structured into five major chapters. In the first, a review of past literature was conducted to have a look at the developments made in solar energy and its different systems. Particularly, we have interest in the solar chimney that is the subject of our study.

The second chapter contains the numerical approach that we used to model the solar chimney. It presents the detailed mathematical model upon which mathematical equations are derived to analyses the design and performance of the solar chimney.

The third chapter presents the choice of the different numerical models responsible for modeling the airflow in the solar chimney. The validations of the numerical results are done with anterior results. In the fourth chapter, we develop a numerical simulation to design our prototype of solar chimney. Particularly, we focus on the study of the effect of the geometrical parameters of the chimney. The numerical results consist of five main parts. It consist also on the study of the solar chimney structure. The validation of the numerical results is done with our experimental results.

In the last chapter, we develop an experimental study for testing the performance of our prototype with different climate conditions in our country.

Finally, we present a general conclusion of this work and the outlook suggested by this study.

CHAPTER 1

Bibliographic Study

1. INTRODUCTION

Renewable energy sources are those that do not rely on stored energy resources. Various forms of renewable energy are currently used for the generation of electricity. As with most industries, the relative cost of a product becomes less expensive as technologies improve and product knowledge increases, with the renewable energy industry being no exception. Using renewable resources, such as the sun, would provide almost unconditional access to energy without depleting the world's natural resources. In this chapter, we are interested in the study of solar energy and its different technologies, particularly to the solar chimney system.

2. SOLAR ENERGY

The sun is the most plentiful energy source for the earth. It is an average star, is a fusion reactor that has been burning over 4 billion years (Askari *et al.*, 2015). The sun emits as much energy to the earth as it is used by the entire population of the planet. In light of the increasing scarcity of fossil fuels, the future of energy supply lies with renewable energy sources and a modern approach to using them. Solar radiation is a free energy source, abundant and renewable. Clean and inexhaustible solar energy helps protect the environment and preserve energy resources, without producing waste or emissions. A global change in energy sources is coming, especially about achieving a healthy balance between economic growth and ecological responsibility for us all, and planet Earth. The future lies in renewable energy sources and modern methods of obtaining them. Innovative and user-friendly solutions are needed to ensure our quality of life (Fig. **1.1**).

The IEA Solar PACES programmer, the European Solar Thermal Electricity Association and Greenpeace estimated global CSP capacity by 2050 at 1 500 GW, with a yearly output of 7 800 TWh, or 21% of the estimated electricity consumption in ETP2010 BLUE Hi-Ren Scenario. In regions with favorable solar resource, the proportion would be much larger. For example, the German

Aerospace Center (DLR), in a detailed study of the renewable energy potential of the MENA (the Middle East and North Africa region) region plus South European countries, estimated that concentrator solar power plants could provide half the electricity consumption around the Mediterranean Sea by 2050 (Fig. **1.1**). We, based on a study by Price Water House Coopers, Europe, and North Africa together could, by 2050 produce all their electricity from renewables if their respective grids are sufficiently interconnected. While North Africa would consume one-quarter of the total, it would produce 40% of it, mostly from onshore wind and solar power (DLR, 2005).

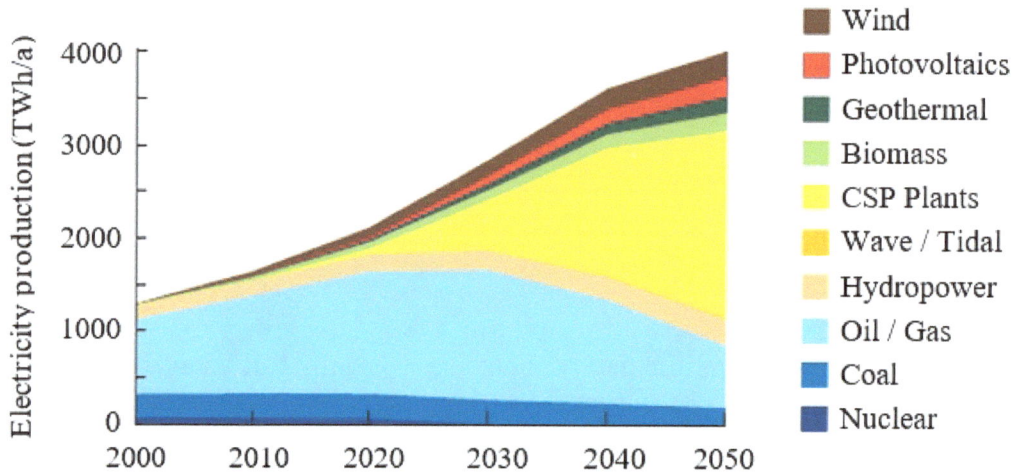

Fig. (1.1). Electricity generation from 2000 to 2050 in 2050 in all MENA region and South-European countries (DLR, 2005).

2.1. Solar Spectrum

The major part of the electromagnetic radiations emitted by the sun is not visible with the naked eye. Fig. (**1.2**) presents a spectrum of the electromagnetic radiations emitted by the sun by including the visible light waves. The naked eye perceives only the rays of which the wavelength lies between 400 and 700 nanometers, which correspond to the cosmic rays. Radiations lower wavelength are called waves decametric, or more usually ultraviolet, and radiations higher wavelength are called infra-red raises. These are the latter, which are responsible for the greenhouse effect that we will be defining later.

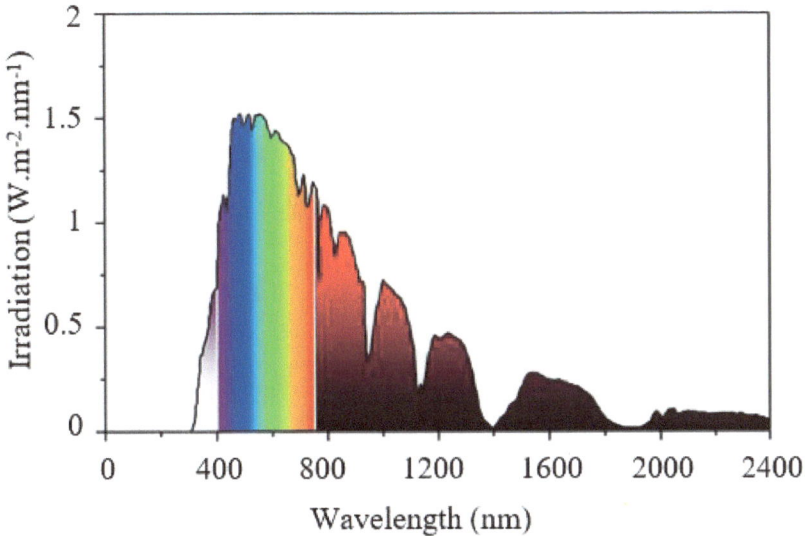

Fig. (1.2). Solar spectrum.

2.2. Irradiation Areas

Geography plays an important role in determining what forms of renewable energy will be the most useful. Solar energy is the primary source of electricity for the third world African countries. These countries use solar energy in isolated regions and cities to harness the sun's energy (Austin *et al.*, 2007). Solar systems are suitable to be applied in arid and semi-arid areas and are an advanced way to generate electricity from solar radiation, average height radiation. Average Horizontal irradiation is another term for the total radiation: the sum of the direct normal irradiance and diffuse horizontal irradiance. Fig. (**1.3**) illustrates the levels of total radiation in Tunisia and the world receives annually. We show that the average solar horizontal irradiation in Tunisia is higher than in most other countries in the world. For this reason, the solar systems represented an important source of energy in our country; thus the company Tunoor would export solar capacities from Tunisia to Europe when it makes a solar power station in the south Tunisia desert.

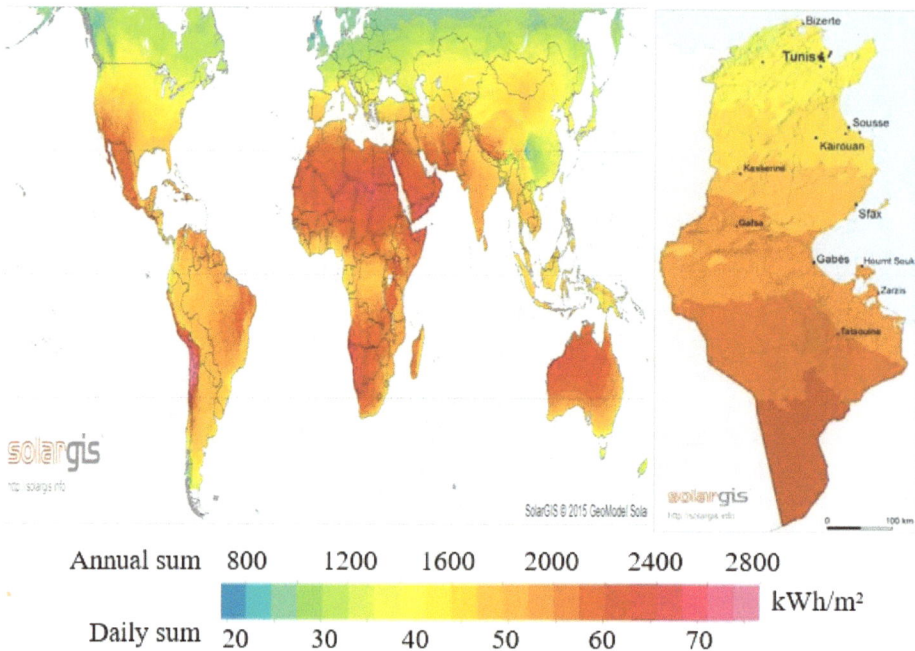

Fig. (1.3). Map of an average annual sum of global horizontal irradiation in the world and Tunisia (Solar GIS).

2.3. Advantages and Disadvantages of Solar Energy

Solar energy is an excellent source of alternative energy because there is no generated pollution while it is used. Moreover, the offer of solar energy to use is unlimited, that means that our dependence with fossil fuels can be reduced. In addition, their costs have not been related to the use of solar energy only but the manufacturing cost of the components, the purchase and the installation of the material are also included. After the initial investment, there are no additional costs associated with its use. In addition, solar installations are flexible. Thus, it is rather easy to increase or to decrease the size of the installation. For example, a solar electric system installed in a room could potentially eliminate 18 tons of polluting gas emission for the purpose of greenhouse in the environment each year. A solar installation can be established anywhere as soon as there is sufficient sun exposure. It is thus a real advantage for the very isolated places, which have access to electricity. Moreover, the use of this energy is a quiet process. No Sound and the harmful effect. One of the major disadvantages is to produce a great quantity of energy that requires a significant installation and sufficient space to install the photovoltaic panels or concentrators. It is thus a considerable constraint for the industrial facilities. Moreover, it is expensive and requires a constant and very intense sunning for the regular commercial practice. Otherwise, much from places in the world do not profit from sufficient sunning to ensure the profitability of the installation.

3. DIFFERENT TYPES OF SOLAR SYSTEMS

The following section lists some of the various types of solar energy sources currently in use.

Today, there are three families of processes for generating electricity using solar energy: either using photovoltaic systems to converting solar energy directly into electrical energy or using thermal systems to converting solar energy into heat and thereafter, converting the heat into electrical energy by thermodynamic systems (Fig. **1.4**).

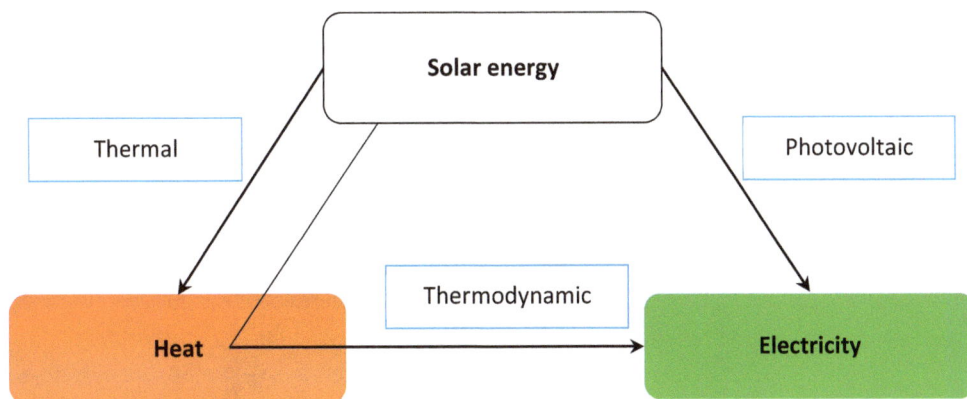

Fig. (1.4). Different types of solar systems.

3.1. System Photovoltaic

The "photovoltaic" term indicated the physical phenomenon overdraft by Alexander Edmond Becquerel in 1839, or the associated technique. The goal of this technique is to convert the energy of the Sun directly into electricity within solar panels. Solar panels produce electricity through individual photovoltaic cells connected in series. This form of energy collection is viable in regions of the world where the sun is plentiful and can be used on houses to supplement the rising cost of electricity from a power grid. To convert the sun's energy, the cells capture photons to create free electrons that flow across the cells to produce usable current (Penick *et al.*, 2007). The efficiency of the panel is determined by the semiconductor material that the cells are made from as well as the process used to construct the cells. Solar panels come in three types: amorphous, monocrystalline, and polycrystalline (Ventre *et al.*, 2000). The more efficient the material the panel is constructed from, the greater the cost. To maximize results, many features can be used to control the output of the photovoltaic panels. The power needs to determine what components are used to produce the desired voltage and current for the project such as converters, solar trackers, and the size of the panel. Converters transform the variable output from solar panels to constant voltages to maximize the continuous supply of usable power for either present needs or stored for future use. The output power of the panel is affected by many variables that continually change throughout the day. Since solar energy is only produced during the day, it requires an energy storage application by either a battery or connecting to the power grid to provide power during the night (Fig. **1.5**).

Fig. (1.5). Solar photovoltaic system.

3.2. Thermal System

Thermal solar energy consists of using the heat of the solar radiation is of direct use of heat: solar water-heaters and heating, solar cookers and steam driers are of indirect use, heat being useful for another use: solar cooling, desalination, *etc*. It is used mainly for the heating of water or the buildings. One uses for that of the thermal sensors. There are several categories, but the principle is the same one:

the thermal sensor absorbs the solar photons and transforms them into heat. This one is transmitted to a liquid or a gas, which transports it (the gas is then known as "coolant") towards a storage tank of heat (Fig. **1.6**).

Fig. (1.6). Solar Hot Water Heating System.

3.3. Thermodynamic System

The solar power stations thermodynamics contain the whole of the techniques, which makes it possible to transform the energy radiated by the sun into heat at high temperature, then to convert this heat into electric power. According to the model of concentration of the solar radiation, there are of configuration different for the thermodynamic solar power stations whose applications are varied: production of electricity, production of vapor for industrial processes or solar supplement for the installations using of the fuels biomasses or fossils.

A thermodynamic system contains five types that can be either a concentration system or a non-concentration system:

- Power stations with parabolic trough collectors (concentration system),
- Linear Fresnel (concentration system),
- Central tower power plants (concentration system),
- Dish Systems (concentration system),
- Solar chimney power plant (non-concentration system),

3.3.1. Parabolic trough systems

The parabolic trough power station with a collector is composed of parallel alignments of long hemicylindrical mirrors, which turn around a horizontal axis to follow the race of the sun (Fig. **1.7**). The solar rays are concentrated on a horizontal tube, where a coolant circulates whose temperature in general reaches 400 °C. This energy is transferred to a circuit from water, the vapor then produced actuates a turbine which produces electricity.

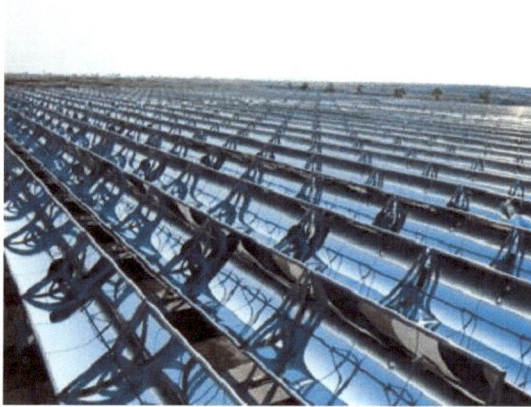

Fig. (1.7). Parabolic trough systems.

3.3.2. Solar Power Stations with Fresnel Mirror

This type of power station based on the principle of the concentrator of Fresnel. Each mirror can swivel to follow the race of the sun of which the goal to concentrate the solar rays towards the fixed linear tubes of reception. In these tubes, a coolant circulates which can be vaporized then overheated at 500 °C that the produced vapor makes it possible to engage a turbine to produce electricity (Fig. **1.8**).

Fig. (1.8). Solar power stations with the Fresnel mirror.

3.3.3. Solar Tower Power Plants

The solar power stations with turn are made up of many mirrors (heliostats) follow the sun concentrating the solar rays towards a boiler located at the top of a tower (Fig. **1.9**). The fluid is then directed either towards storage or towards the exchanger where heat is transmitted to air or water, at a temperature which can vary from 600 with 1000 °C. The generated vapor actuates turbines to produce electricity. The power conversion technology for troughs and power towers is a conventional steam Rankin power cycle.

Fig. (1.9). Solar tower power plants.

3.3.4. Dish Systems

The parabolic mirror reflects the rays of the sun towards a point of convergence (Fig. **1.10**). The receiver in question is a Sterling engine which functions thanks to the rise in temperature and pressure of a gas contained in a closed enclosure. This engine converts thermal solar energy into mechanical energy and then into electricity. Throughout the day, the base of the parabola is directed automatically with the sun to follow its race and thus to benefit from a maximum sunning. Solar dishes systems are efficient at collecting solar energy at very high temperatures, causing the high concentration ratios achievable with parabolic dishes and the small size of the receiver.

Fig. (1.10). Dish sterling system.

3.3.5. Limitation of Photovoltaics and Thermodynamic System

Despite, the promoters of the technique themselves recognize that these processes are only viable in the sunniest regions of the world, receiving at least 1,900 kWh per m^2 per year, *i.e.* tinted areas. in yellow on the map below, which are located exclusively south of the 40th parallel. We will easily recognize the main deserts of the planet, which is no coincidence.

Given the large areas required by these installations to obtain significant installed power, the possible sites are precisely limited to desert or semi-desert areas, or in any case sparsely populated. In addition, these areas should, if possible, avoid the close presence of mountain ranges which would provide shade at the start and end of the day. In the European Union, the only areas that meet these criteria are in southern Spain and the Mediterranean islands.

-They contain toxic products and the recycling industry is not yet in existence;
-The investment cost is currently very high.
-The electrical efficiency decreases over time (20% less after 20 years).
-When the storage of electrical energy by batteries is necessary, the cost of the photovoltaic system increases.
-As in the case of a solar thermal collector, the production of solar energy is only possible when there is the sun.

Despite many favorable characteristics, PV / T systems still suffer from several shortcomings, which hamper the commercialization of this technology.

• Careful design of the PV / T, especially for electrical insulation and efficient heat transfer, is necessary because electrical and thermal devices are combined here to operate simultaneously.
• Poor thermal contact between the PV module and the coolant in the flow channel leads to a temperature difference of around 15 ° C in the case of unglazed PV / T. This is due to the increased resistance to heat transfer between the cell and the absorber interface leading to a low rate of heat dissipation. The search for suitable thermally conductive adhesives is a major problem in PV / T manufacturing.

3.3.6. Solar Chimney Power Plant

In this technology, solar radiations are not concentrated. The non-concentrating technology on the other hand employs collectors that are unable to concentrate the solar radiations *i.e.* they capture both the direct and diffuse solar irradiation. This technology is the goal of our next study in memory.

4. SOLAR CHIMNEY

This study aims to present a comprehensive description of the solar chimney and its application for power generation. This description is the product of bibliographic research and the collection of information available on the subject.

4.1. Description

The solar chimney is a means of electricity production from solar energy. It based on the idea of using natural convection of air heated by solar radiation. A huge greenhouse called "collector" contains and guides the air. This hot air, naturally aspirated through the chimney, is driven by a pressure difference in the chimney system. So traditional chimneys are extremely tall to increase the pressure differential and the air's velocity. The flow is continuously renewed by air located on the outskirts of the greenhouse. Equipped with water balloons that absorb heat at the day to restore at the night, a steady wind then starts up. The kinetic energy of the air is then imposed by a system of turbines and generators. This technology is based on a simple principle: The sunlight penetrates the transparent cover. A faction of incident radiation reaches the ground, which increases its temperature. The heat from the ground is transferred by convection to the air underneath the canopy. This heat transfer is increased due to the greenhouse effect of the transparent roof (Schlaich *et al.*, 1995). The efficiency of the solar chimney power plant is below 2% and depends mainly on the height of the tower. As a result, these power plants can only be constructed on land that is very cheap or free. Such areas are usually situated in desert regions (Chikere *et al.*, 2011) (Fig. **1.11**).

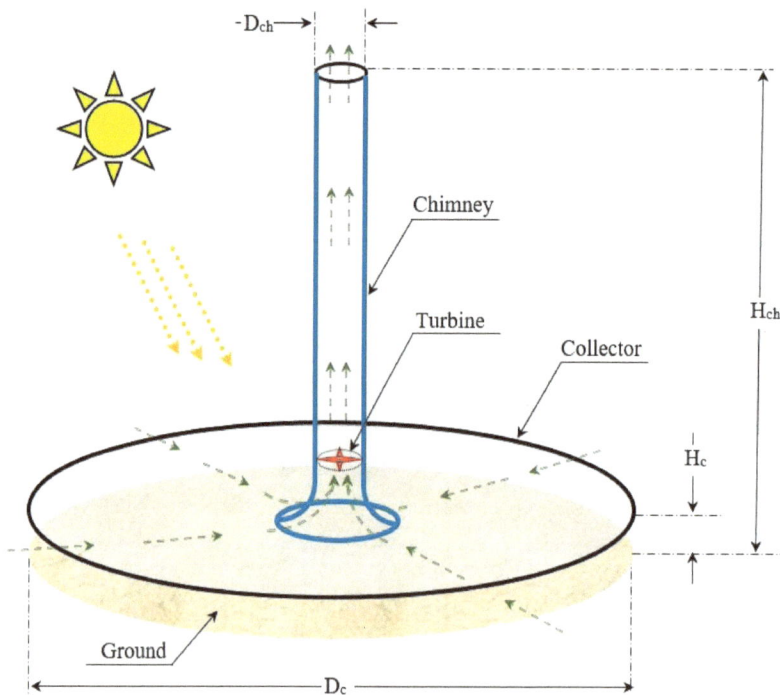

Fig. (1.11). Solar chimney system

4.2. Greenhouse Effect

The term effect of the greenhouse is usually employed to speak about the pollution of the atmosphere by gases called "for the purpose of greenhouse" such as CO2 or methane. However, more generally, the greenhouse effect is a principle used in many cases to imprison heat to preserve or maintain the elements such as the plants. The principle of the greenhouse effect (Fig. **1.12**) most generally uses glass. Indeed, first, the sun emits all rays, which cross a glass surface and arrive at the ground except infrared rays. These rays are then considered on the ground, which returns them after they increased their wavelength. This means that the ground absorbs the rays and emits again when it creates the infrared rays. However, one of the characteristics of glass is to let pass the solar rays but very few infrared rays. Even the clear glazing, are practically opaque with radiations wavelength being in the infrared. Thus, by increasing the wavelength of the rays, those are imprisoned between the ground and glass, which causes a rapid raises temperature and also makes it possible to store heat (Hassanien *et al.*, 2016).

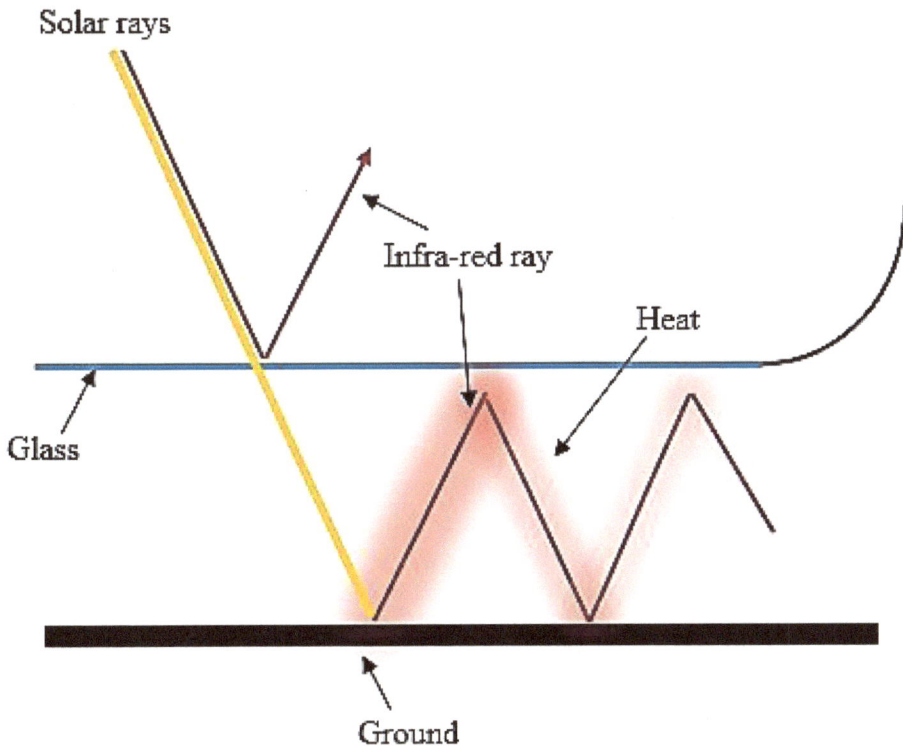

Fig. (1.12). Greenhouse effect.

4.3. Principle

The natural convection created by the collector can transform thermal energy into kinetic energy in the chimney. This kinetic energy will be converted into mechanical energy by means of the turbines. This mechanical energy is transformed finally into electric power thanks to the alternators related to the turbines (Fig. **1.13**).

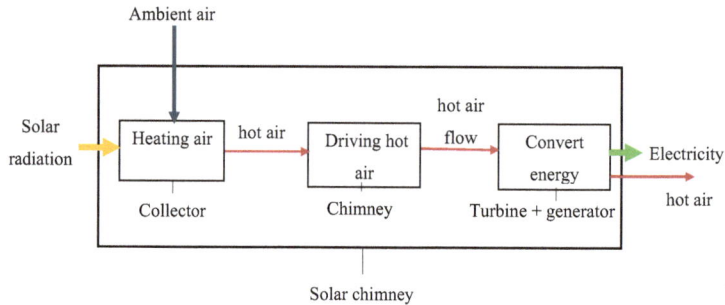

Fig. **(1.13)**. Schematic diagram of solar chimney.

4.4. Mainly Components of the Solar Chimney

4.4.1. Collector

The collector is the part of the solar chimney that produces hot air by the greenhouse effect. It consists of a support matrix, column structure, and a transparent roof. A large air collector is formed when a transparent glass or plastic roof supported above the ground by column structure and support matrix is stretched out horizontally many meters. The height of the roof slowly increases along a radius from the periphery to the center to guide inward airflow with minimum friction losses. This transparent roof admits direct and diffused solar beam to be transmitted and retains long-wave radiation from the heated ground. The greenhouse effect is therefore produced in the collector. The structure of the collector changes to the covering material we used. Since the weight of the collector roof increases the mass of the roof, collectors must have a structure with resistant and attached rods as presented in Fig. **(1.14)**.

Fig. (1.14). Collector structure design options.

4.4.2. Chimney

The chimney is the most important part of the solar tower. It is a pressure tube with low friction loss. It acts as a heat engine which makes it possible to create a gradient of pressure between the airs locates in the hotter bottom (less dense) and the air locates at the exit of the chimney.

However, compared with the collector and the turbines, the chimney efficiency is relatively low, hence the importance of size in its efficiency curves. The chimney should be as tall as possible for example at 1000 m height can be built without difficulty. The material for the chimney construction could be reinforced concrete tubes, steel sheet tubes supported by guy wires, or cable-net construction with a cladding of sheet metal or membranes (Fig. **1.15**).

Schlaich (1995) suggests that reinforced concrete would be a cost-effective way to create a stable tower with a life span of up to 100 years. In other terms, there are two technologies of the chimney with the simple, which we know (Fig. **1.16**).

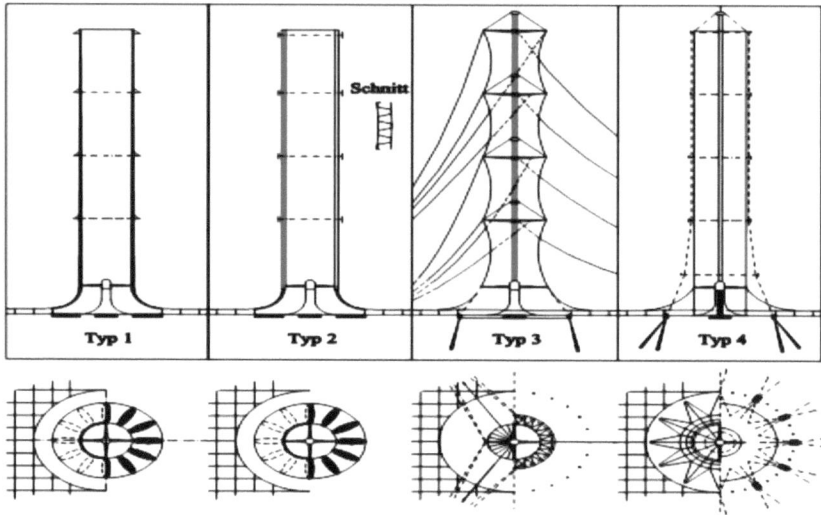

Fig. (1.15). Chimney construction shapes.

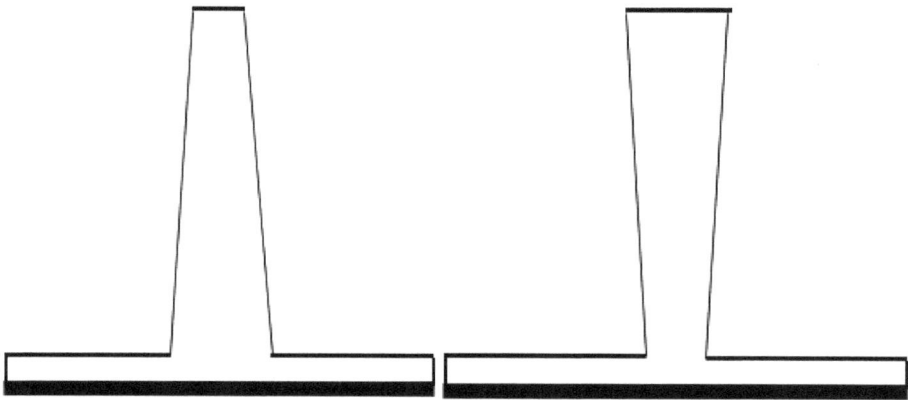

(a) convergent-top chimney.   (b) divergent-top chimney.

Fig. (1.16). Schematic layout of different technology of chimney.

4.4.3. Turbine

The turbine in a solar chimney does not work with a staged velocity like a free-running wind energy converter, but as a closed pressure-staged wind turbogenerator, in which, similarly to a hydroelectric power station, static pressure is converted to rotational energy using a cased turbine in this application installed in a pipe. The energy yield of a cased pressure-staged turbine of this kind is about eight times greater than that of a speed-stepped open-air turbine of the same diameter. The wind turbines used in the solar chimney are ducted and so their maximum achievable theoretical total efficiency can go beyond 59% as stipulated by Betz limit in those of wind generators (Pastohr, 2003).

The turbines are always placed at the base of the chimney where the rate of flow of air is raised, to transform the maximum of energy kinetic into mechanical energy. The generator driven by the turbine transforms the mechanical energy into electric energy. The number of turbines used is either a turbine of which the blades cover the transverse section of the chimney (Fig. **1.17c**) or of small turbines distributed on all the sections of the chimney (Fig. **1.17b**). But, it is also possible to arrange between the absorber and the base of the chimney a great number of small turbines with horizontal axes (Fig. **1.17a**).

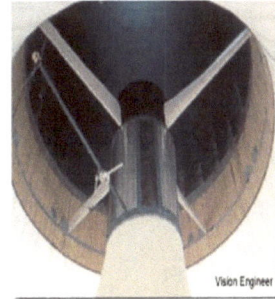

(a) Small turbines with horizontal axes   (b) Small turbines with vertical axes   (c) Large turbine

Fig. (1.17). Different technologies of the turbine.

4.5. Energy Storage

During the day, the solar chimney functions by the solar radiation but the absence of the solar radiation (during the night), will create a stop of the chimney (Fig. **1.18**). For this reason, the storage of energy using the radiation of the day to make the function of the chimney during the night. There are several methods of storage of energy. Pretorius (2004) investigated six different ground types and they include sandstone, granite, limestone, and sand, wet soil, and water. Another type of thermal storage that has attracted several types of research is the use of water-filled black tubes. It is laid down side by side on the black sheeted or sprayed soil under the glass roof collector. They are filled with water once and remain closed thereafter. So that no evaporation can take place. The volume of water in the tubes is selected to correspond to a water layer with a depth of 5 to 20 cm depending on the desired power output. Since the heat transfer between black tubes and water is much larger than that between the black sheet and the soil, even at low water flow speed in the tubes, and since the heat capacity of water (4.2 KJ.kg-1) is much higher than that of soil (0.75 - 0.85 KJ.kg-1). The water inside the tubes stores a part of the solar heat and releases it during the night, when the air in the collector cools down. Kreetz (1997) examines the influence of the groundwater storage on solar chimney power. Fig. (**1.19**) shows that with the water-filled tubes, power can be generated in the night but with a drop in the power generated during the hours of sunshine. Hence we noticed that the possibility of a continuous day and night operation of the solar chimney.

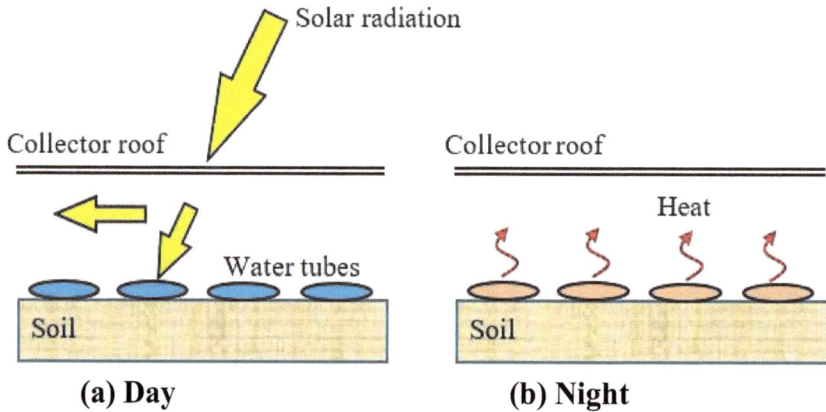

Fig. (1.18). Storage with water fields.

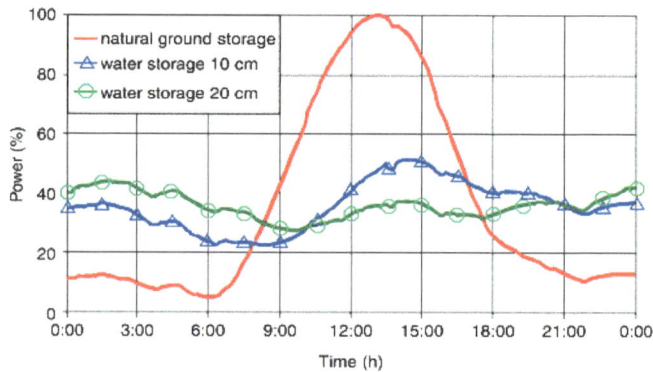

Fig. (1.19). Effect of heat storage underneath the collector roof using water-filled black tubes (Kreetz, 1997).

4.6. Advantages and Disadvantages

The solar chimney has several advantages in comparison with other power production technologies (Schlaich, 1995):

- The collector can use all solar radiation, both direct and diffused, so it can operate in cloudy conditions.
- The solar chimney will operate 24 h only on pure solar energy due to the heat storage system.
- The ground provides natural heat storage.
- Solar chimney power plants do not require cooling water.
- The plant has a long operating life (at least 80 to 100 years).
- Solar chimneys are particularly reliable and not liable to break down.
- The building materials needed for solar chimneys, mainly concrete and glass, are available everywhere in sufficient quantities.
- Solar chimneys can be built in less industrially developed countries even in poor countries, do not need investment in a high-tech manufacturing plant.
- Low maintenance cost.

The main disadvantages of the solar chimney are:

- The initial capital cost is high.
- The construction of the chimney requires enormous amounts of material. These amounts may cause logistical problems belonging to the availability and transportation of materials.
- The power output is not constant throughout the day or year.
- Negative visual impact (some see a degradation of the landscape).

A comparison with price and cost with another source renewable energy showed in Table **1.1** is a comparison with the photovoltaic and the nuclear power which, indicates an advantage to the profit of the nuclear power and another level of comparison with the photovoltaic one and the nuclear power on the price plan of produced kWh indicates an advantage to the profit of the solar chimneys (kaltschmitt *et al.*, 2007).

Table 1.1. Comparison with price and cost with another source of renewable energy.

	Solar chimney		Photovoltaic		Nuclear	
	Australia project	Namibia project	Andasol	Mildura	Civaux	European Pressurized Reactor
Power (MW)	200	400	50	154	1450	1600
Price of kWh (€)	7	2	27	9	11 € hundred, without counting the price waste disposal, insurances accidents	

4.7. The Efficiency of Solar Chimney

The influence of the essential parameters on the power output of a solar tower is presented here in a simplified form. The total effectiveness of the solar chimney is determined by Schlaich *et al.* 1995) that the overall efficiency is the multiplication of collector, chimney turbine, and generator:

$$\eta = \eta_c \ \eta_{ch} \ \eta_t \ \eta_g \tag{1-1}$$

Where η_c is the collector efficiency, η_{ch} is the chimney efficiency, η_t is the turbine efficiency and η_g is the generator efficiency.

4.7.1. Collector Efficiency

Collector efficiency is improved as a rise in temperature decreases. Thus, a solar chimney collector is economic, simple in operation, and has a high-energy efficiency level. A solar chimney collector converts available solar radiation G onto the collector surface A_c into heat output. Collector efficiency η_c can be expressed as a ratio of the specific rate of heat outflow Q and the solar energy input Q_s (Schlaich *et al.*,1995).

The solar energy input Q_s can be written as the product of the global radiation G and collector area A_c.

$$\eta_c = \frac{Q}{Q_s} = \frac{Q}{A_c \cdot G} \tag{1-2}$$

$$Q = \dot{m} \ c_p \ \Delta T \tag{1-3}$$

$$\dot{m} = \rho_c \ A_c \ v_c \tag{1-4}$$

Where:

c_p : specific heat capacity of the air,

ΔT : air temperature increase between the collector inflow and outflow,

$\dot{m}$: mass flow in the chimney,

ρ_c : specific density of air at temperature To + ΔT at collector outflow/chimney inflow,

A_c : Chimney cross-section area,

V_c : Air speed at collector outflow/chimney inflow,

Therefore, the collector efficiency:

$$\eta_c = \frac{\rho_c\, A_c\ v_c\, c_p\, \Delta T}{A_c\, . G} \qquad (1\text{-}5)$$

Some theoretical investigations (Schlaich *et al.*, 1995) have considered that the energy equation balance of the collector is given by the specific rate of heat outflow in the collector (W). It can be estimated from the following expressions (Fig. **1.20**):

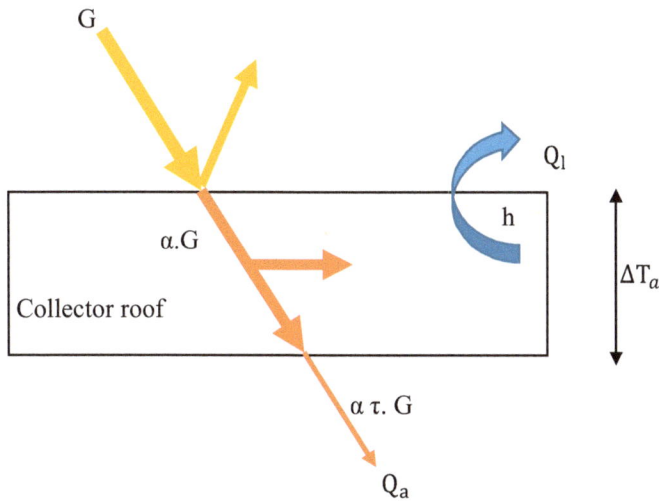

Fig. (1.20). Collector Energy Balance.

$$Q \;=\; Q_a - Q_l \qquad (1\text{-}6)$$

$$Q \;=\; A_c(\alpha\tau.\, G - h.\, \Delta T_a) \qquad (1\text{-}7)$$

Where α is the effective absorption coefficient of the collector, transmittance coefficient of the collector, and h is the loss correction value (in $W.m^{-2}.K^{-1}$), allowing for emission and convection loss ΔT_a. is the average difference in temperature under the collector and the ambient. It can be estimated by the difference of temperature between the outlet collector and the ambient divided by two.

$$\Delta T_a = \frac{\Delta T}{2} \tag{1-8}$$

Therefore, the collector efficiency is given by another formula:

$$\eta_c = \alpha\tau - h \frac{\Delta T}{2G} \tag{1-9}$$

4.7.2. The efficiency of the chimney

The density difference of the air caused by the temperature rise in the collector works as a driving force. The lighter column of air in the tower is connected with the surrounding atmosphere at the base inside the collector and the top of the tower, and thus acquires lift. A pressure difference ΔP_t is produced between the tower base (collector outlet) and the ambient. Moreover this pressure difference increases with chimney height.

$$\eta_{ch} = \frac{P_t}{Q} \tag{1-10}$$

$$dP = g.\rho.dz \tag{1-11}$$

$$\Delta p_t = \int_0^H g.(\rho - \rho_c)dz \tag{1-12}$$

ΔP_t consists of two components, the static pressure difference drops at the turbine and the dynamic component describing the kinetic energy of the airflow. Assuming the pressure loss due to friction is negligible then:

$$\Delta p_t = \Delta p_t + \Delta p_d \tag{1-13}$$

Without the turbine (x=0)

$$\Delta p_d = \frac{1}{2}\rho_c . v_{c,max}^2 \tag{1-14}$$

With the total pressure difference and the volume flow of the air at $\Delta P_t =$

0(Without the turbine), the power contained in the flow is defined as follows:

$$\Delta p_t = \frac{1}{2} \cdot \rho_c \cdot v_{c,max}{}^2 \qquad (1\text{-}15)$$

Using the Boussinesq approximation (Unger, 1988), the speed reached by free convection currents can be expressed as:

$$v_{c,max} = \sqrt{2gH \frac{\Delta T}{T_0}} \qquad (1\text{-}16)$$

Thus:

$$\Delta p_t = \rho_c\, g\, H\, \frac{\Delta T}{T_0} \qquad (1\text{-}17)$$

Where:

T0: the ambient temperature at ground level,

ΔT: Temperature rises between collector inflow and collector outflow /chimney inflow,

Therefore, the Efficiency of the chimney is given by:

$$\eta_{ch} = \frac{\frac{1}{2} m\ v_{c,max}{}^2}{m\ c_p\ \Delta T} = \frac{g\,H}{c_p T_0} \qquad (1\text{-}18)$$

This basic simplified explanation one of the basic characteristics of the solar chimney, which is that the chimney efficiency is fundamentally dependent only on chimney height. Flow speed and temperature rise in the collector do not come into it.

4.7.3. The Eefficiency of the Turbine and Generator

The ratio of pressure drop across the turbine determines the available plant power output.

$$\eta_t = \frac{\Delta p_t}{\Delta p_{tot}} = x \tag{1-19}$$

The pressure drop across the turbine can be expressed as a function of the total pressure difference:

$$\Delta p_t = x \, \Delta p_{tot} = x \, (\Delta p_t + \Delta p_d) \tag{1-20}$$

With the turbine:

$$\Delta p_d = \frac{1}{2} \rho_c \cdot v_t^2 \tag{1-21}$$

$$v_t = v_{c,max} \sqrt{1 - x} \tag{1-22}$$

The optimal x for the maximum power extraction can be obtained by assuming that v_t and ΔP_{tot} are not functions of x and solving $\partial/\partial x \, (P_t) = 0$. The result for the optimal value obtained by Koonsrisuk et al (2012) is equal to ⅔.

4.7.4. Power

Generally, the theoretical electric power output P_e of the solar chimney can be calculated as the solar input Q_s multiplied by the respective efficiencies of collector, chimney, turbine, and generator.

Without the turbine, power contained in the flow:

$$P = \eta_{ch} \, \eta_c \, Q_s = \rho_c \, A_c \, v_c \, \frac{\Delta T}{T_0} \, g \, H \tag{1-23}$$

With a turbine, mechanical power produced by the turbine is written as follow:

$$P = \eta_{ch} \, \eta_c \, \eta_{tg} \, Q_s = x \, \eta_g \, A_c \, v_c \, \rho_c \, g \, H \, \frac{\Delta T}{T_0} = A_c \, v_c \, \Delta p_t \tag{1-24}$$

Electric power is obtained while multiplying Mechanical power by η_g:

$$P_e = U\,I = x\,\eta_g\,A_c\,v_c\,\rho_c\,g\,H\,\frac{\Delta T}{T_0} = x\,\eta_g\,\eta_c\,\frac{g\,H}{c_p T_0}\,A_c\,G \qquad \textbf{(1-25)}$$

5. A SOLAR CHIMNEY TIMELINE

A solar chimney (also called solar updraft tower power plant, SUTPP) is a kind of device that produces buoyancy to drive air to ascend for electricity generation (Schlaich., 1995). The fact that more than century scientists and researchers are addressing the topic of this technology. Around 1500, Leonardo da Vinci created the earliest system, which uses hot air rising in a chimney to drive an apparatus. One of his sketches depicts a roasting spit driven by a turbine located in the chimney above a fireplace as shown in Fig. (**1.21.a**) (Calder, 1970).

In 1903, the Spanish engineer Isodoro Cabanyes proposed the first idea of generating electricity from the solar chimney (Fig. **1.21.b**).

In 1926, Engineer Bernard Dubos proposed to the French Academy of Sciences the construction of a Solar Aero-Electric Power Plant in North Africa with its solar chimney on the slope of the high height mountain after observing several sand whirls in the southern Sahara as presented in Fig. (**1.21.c**) (Hamilton, 2011).

In 1931, the German science writer Hans Gunther wrote a book that showed the first contemporary citing of a solar chimney based on the system proposed by Dubos. He proposed a design in the 25 August 1903 issue of Energeia Electricall, entitled Projector de motor solar. In this bizarre contraption, a collector resembling a large skirt heats the air and carries it upwards towards a pentagonal fan inside a rectangular brick structure vaguely resembling a fireplace (without a fire). The heated air makes the fan spin and generates electricity, before it escapes up a 63.87 m tall chimney, cools, and joins the atmosphere (OMRI *et al.*, 2013).

In 1956, in France, the military Edgard Nazare, after he scientifically measured the sands of the Sahara swirls, deposed in Algiers a first patent on what he called "the tower to depression." This patent was reintroduced in Paris on 3 August 1964. It was artificially generating a swirling atmospheric rise in a kind of round-shaped Laval nozzle. However, Nazare died in Paris in September 1998, without having been able to build a tower of 300 m close to his heart. In 1975, the Canadian engineer Louis Michaud published a project called Vortex Power Station: and it is very similar to Nazare project. However, Michaud proposed to initiate the vortex phenomenon with burners at the bottom of the tower. While Nazare thought that the phenomenon could be initiated by the simple stack effect (Fig. **1.21.d**).

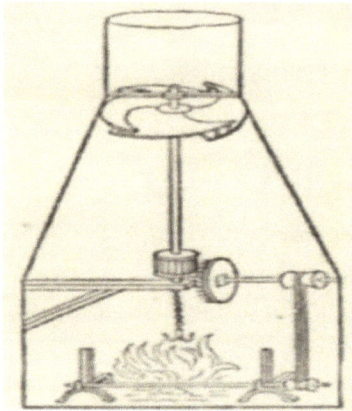

(a) System proposed by
Leonardo da Vinci

(b) System proposed by
Isodoro Cabanyes

(c) System proposed by
Bernard Dubos

(d) System proposed by
Edgard Nazare

Fig. (1.21). First systems of solar chimney in the world.

In 1982, Schlaich together with his colleagues built the first pilot SUTPP prototype in Manzanares Spain (Fig. **1.22**). The pilot prototype had an SUT 194.6 m high and a collector 122 m in radius. The prototype operated with a peak power of about 50 kW for seven years from 1983 to 1989 (Schlaich, 1995). The successful operation of the prototype demonstrated the feasibility and reliability of SUTPP technology. Since then, many researchers have shown a strong interest in it and extensively studied the potential of SUTPP technology all over the world (Zhou *et al.*, 2010). To generate electricity economically, a large-area collector and a high SUT are needed for a SUTPP. In the past decades, several experimental models were successively designed, built up, and tested, whose structures differed from each other. The first 50 kW plant prototype built in Manzanares in a view to measurement, which had 194.6 m high, 0.00125 m-thickness metallic wall SC guyed, and a PVC roof-covered collector 122 m in radius. The main structure parameters and technical data of Manzanares plant are illustrated in Table **1.2**.

Table 1.2. Main structure parameters and technical data of Manzanares plant.

Item	Value
Chimney height (m)	194.6
Chimney radius (m)	5.08
Mean collector radius (m)	122
Mean roof height (m)	1.85
Number of turbine blades	4
Turbine blade profile	FX W-151-A
Blade tip speed to air transport velocity ratio (m)	10:1
Operation modes	Stand-alone or grid connected mode
Typical collector air temperature rise (K)	T=20
Nominal power output (kW)	50
Collector covered with a plastic membrane (m^2)	40,000
Collector covered with glass (m^2)	6000

(a) Global system   (b) Collector   (c) Turbine

Fig. (1.22). Solar chimney prototype in Manzanares.

5.1. Mainly Current Projects

Based on the Manzanares prototype, many research works are being carried out which involves the construction of different projects to investigate the potential of solar chimneys power all over the world. Fig. (**1.23**) shows the largest projects in the world with its power. A few of those projects will be outlined.

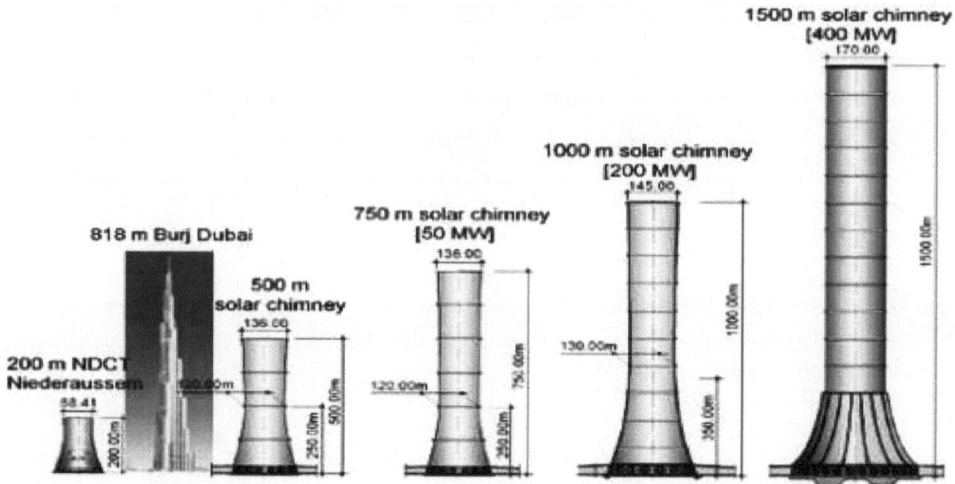

Fig. (1.23). Solar chimney in the world.

In 2001, a company called Environ Mission initiated a plan to build a 200 megawatts solar chimney in southwest Australia that could generate 4000 times more power than the Manzanares system (Fig. **1.24**). The plant having seven-kilometer roof diameter and 1 km chimney height, and a 3-meters distance at the outer periphery and 25 m distance at the inner periphery of the solar collector roof and which it allows to sucked hot air through 32 turbines which generate power 24 hours a day.

In 2008, the Namibian government proposed the idea of constructing a solar chimney called the Green tower (Fig. **1.25**). This solar chimney is the tallest structure on land, it will be 1.5 kilometers in height and 280 meters in diameter and will function by forcing heated air through a shaft lined with wind turbines, producing 400 MW of electricity. The area of the greenhouse is 38.5 km^2 (7 km diameter) and a total estimated cost of 01 billion dollars.

Fig. (1.24). Environ mission power plant scheme.

Fig. (1.25). Namibian solar chimney.

5.2. Floating Solar Chimney Technology

To increase the efficiency of solar chimney power plants and to decrease their cost, Papageorgiou, proposed to build higher and less costly solar chimneys, called floating solar chimney, as a lighter than air structure, can be raised anywhere and its cost is as low as 2% of the cost of the respective concrete chimney (Papageorgiou, 2006). The major advantage behind this concept is that it seeks to replace conventional concrete solar chimneys due to its lower construction cost as compared to that of concrete. The floating solar chimney will have its chimney made with a flexible material and will float on air with the help of a lighter gas like helium. The chimney essentially has a heavy base and the walls are filled with lighter gas. The support rings allow air to enter and pass through them freely. So that the chimney does not yield under wind pressure (Fig. **1.26**).

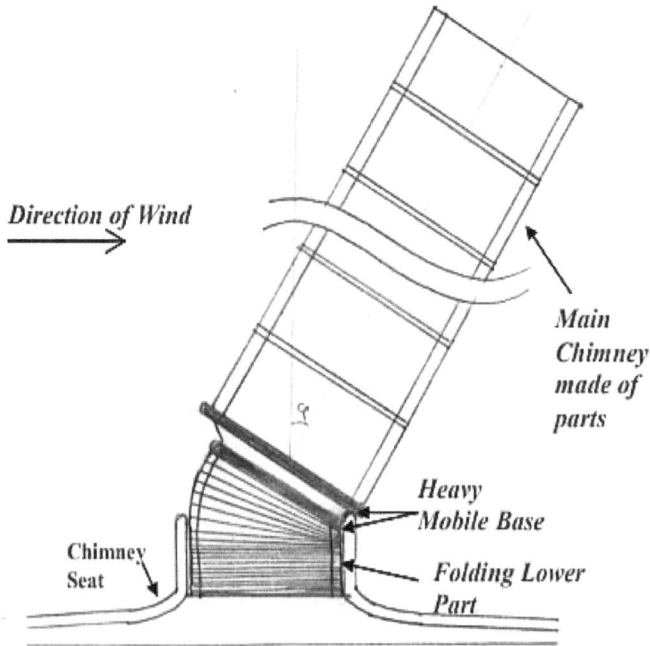

Fig. (1.26). Floating solar chimney technology.

5.3. Experimental Prototypes

The power output profile correlates closely with solar insolation profile during the day time for this prototype plant without additional storage system, while, there is still an updraft during night time due to thermal storage capacity of natural soil, which can be used to produce power during some hours of the night (Haaf *et al.*, 1984). In 1983, Krisst built a courtyard SC power setup with 10 W power output.

The collector base diameter and SC height were 6 m and 10 m, respectively. In 1985, a micro-scale model with an SC 2 m high and 7 cm diameter and a 9 m^2 collector was built by Kulunk (1985) in Turkey. In 1997, an SC power setup was built by

Pasurmarthi and Sherif in Florida, which had a Lexan roof covered collector 9.15 m in diameter and a 7.92 m high SC with its diameter gradually decreasing from 2.44 m at the inlet to 0.61 m at the top (Sherif, 1995). An aluminum plate absorber was laid down on the ground under the collector roof (Type I configuration). Two enhancements were tried on the type I configuration collector to increase the power output. The type I configuration collector base was extended to form type II configuration collector 18.3 m in diameter. Black visqueen and clear visqueen with transparency of about 60% were respectively used as the absorber and the roof for the extended parts of the type II configuration collector. An intermediate canvas absorber was introduced between the roof and the aluminum plate absorber inside the type II configuration collector to improve the conversion efficiency of collector and form type III configuration collector. For the three configuration collectors at insolation of 650 W.m-2. The type I configuration collector air temperature rise was about 15 °C, whereas the type II and type III collectors were able to raise the temperature by 25 and 28 °C, respectively. These showed the type I configuration collector was not very effective, and in type II and type III collectors, the temperature variation in the extended section was almost the same. Whereas, in the Lexan roof covered collector section, there was a marginal improvement in the type III collector compared to the type II collector. In the type III collector, air flowed on either side of the extended canvas absorber, thus inducing an increase in mass flow rate, and hence power output. A pilot SC setup consisting of an air collector 10 m in diameter and an 8 might SC was built in Wuhan, China (Zhou *et al.*, 2008) in 2002 and re-built several times. For up-to-date structure, the collector roof and the SC were made of glass 4.8 mm in thickness and PVC, respectively. The temperature difference between the collector outlet and the ambient usually could reach 24.1 °C. An interesting phenomenon was found that air temperature inversion appeared in the latter SC after sunrise both on a cool day and warm day. The air temperature inversion was formed by the increasing process of insolation from the minimum and cleared up sometime later when the absorber bed was heated to a high enough temperature to let airflow break through the temperature inversion layer and normally flow out from the SC outlet. Based on the need for plans for long-term energy strategies, Botswana's Ministry of Science and Technology designed and built a pilot SC power setup for research (Ketlogetswe *et al.*, 2008) SC was manufactured from glass reinforced polyester material, which had an inner diameter of 2 m and a height of 22 m. The collector roof was made of a 5 mm thick clear glass supported by a steel framework. The collection area reached approximately 160 m². The absorber under the roof was made of two layers of compacted soil approximately 10 mm thick and a layer of crushed stones. The layer of crushed stones was spread on the top surface of the compacted soil layer. The schematic diagrams of other testing models are shown in Fig. (**1.27**).

(a) Sherif et al. (1995).   (b) Zhou et al. (2007 ; 2008).

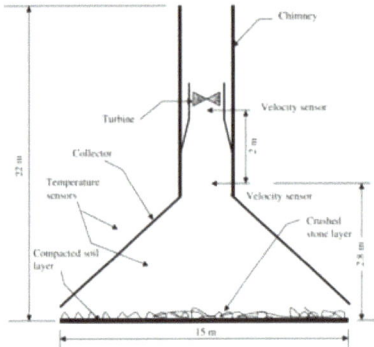

(c) Ketlogetswe et al. (2008).   (d) Maia et al. (2009).

(e) Koyun et al. (2007).   (f) Ketlogetswe et al. (2008).

Fig. (1.27). Schematic diagrams of other testing models.

CONCLUSION

From the above discussion, this chapter would like to draw that solar energy is a good source of renewable energy, but in this technology, we noticed that the solar chimney is the more interesting solar technology. Hence, due to various advantages now, most of the countries are attracting the generation of power by using solar chimney systems. The purpose of this chapter is to define and study the different parts of the solar chimney to pave the study of airflow through a solar chimney.

CHAPTER 2

Numerical Approach

1. INTRODUCTION

Modeling is mainly the basis of engineering and it is a suitable simplification of reality. The displaying expertise is to detect the fitting degree of disentanglement, recognize significant highlights from those that are insignificant in a specific application, and use designing judgment. This chapter presents a complete model specification of the solar chimney power plant on which all simulations and results are based and devoted to the presentation of the physical and numerical models by using in the CFD code ANSYS Fluent 17.0. The following will discuss the equations and models incorporated by ANSYS Fluent to model a solar chimney power plant. The governing equations will be presented, which describe the coupling of temperature fields, velocity, and pressure from the conservation equations. Then, the discretization methods and meshing used will be examined. Finally, the models for turbulence and radiation will be explored.

2. MATHEMATICAL FORMULATION

2.1. Governing Conservation Equations

The equations governing the flow of air in a solar chimney are the Navier Stokes equations. ANSYS Fluent 17.0 employs the fundamental three-dimensional fluid mechanic equations for conservation of mass (The continuity equation), momentum, and energy to calculate fluid properties.

The continuity equation expresses the law of conservation of mass for given volume control:

$$\frac{d\rho}{dt} + \vec{\nabla} \cdot (\rho \vec{V}) = 0 \qquad \text{(2-1)}$$

Where:

$$\vec{\nabla} = \left(\frac{\partial}{\partial x}, \frac{\partial}{\partial y} \right) \qquad \text{(2-2)}$$

$$\text{div } \vec{v} = \nabla . \, \vec{v} \tag{2-3}$$

t presents the time, ρ is the density, and $\vec{V}$ is the velocity vector.

$$\frac{\partial}{\partial x_i} (\rho u_i) = 0 \tag{2-4}$$

The momentum equations are written as follows:

$$\frac{d\rho\vec{V}}{dt} + \left(\rho \vec{V}.\vec{\nabla} \right).\vec{V} = -\vec{\nabla}p + \vec{\nabla}.\bar{\bar{\tau}} + \rho . \, \vec{g} \tag{2-5}$$

Where p is the static pressure.

The stress tensor $\bar{\bar{\tau}}$ is expressed as:

$$\bar{\bar{\tau}} = 2\mu\bar{\bar{D}} - \vartheta(\overrightarrow{\nabla.V}).\bar{\bar{I}} \tag{2-6}$$

Where:

$$\bar{\bar{D}} = \frac{1}{2}(\bar{\bar{\nabla}}.\vec{V} + \bar{\bar{\nabla}}^t.\overrightarrow{V}) \tag{2-7}$$

Thus,

$$\bar{\bar{\tau}} = \mu((\bar{\bar{\nabla}}.\vec{V} + \bar{\bar{\nabla}}^t.\overrightarrow{V}) - \frac{2}{3}(\overrightarrow{\nabla.V}).\bar{\bar{I}} \tag{2-8}$$

Where $\bar{\bar{I}}$ the identity matrix and μ is the dynamic viscosity.

The final term in the momentum equation represents the buoyancy force.

$$\frac{\partial}{\partial x_j} (\rho u_i u_j) = -\frac{\partial p}{\partial x_i} + \frac{\partial \tau_{ij}}{\partial x_j} + \rho g_i \tag{2-9}$$

Basing on the first principle of thermodynamics, we can express the conservation equation of energy. This principle connects the various forms of energy, that is to say:

$$\frac{d\rho E}{dt} + \vec{\nabla}(\rho E \vec{V}) = \vec{\nabla}.(\overline{\overline{\tau}}.\vec{V}) + \rho \vec{f}.\vec{V} - \vec{\nabla}.\vec{q} - \vec{\nabla}.p\vec{V} + r \qquad (2\text{-}10)$$

Where:

E is the total of the potential, kinetic, and internal energy in the system.

The internal energy, h, is the enthalpy of the fluid and is expressed as:

$$h = \int C_p dT \qquad (2\text{-}11)$$

Where:

C_p: is the constant pressure specific heat

r: is an additional energy source term.

$$\frac{\partial}{\partial x_j}(\rho C_p u_j T) = \frac{\partial}{\partial x_j}\left(k\frac{\partial T}{\partial x_j}\right) + \tau_{ij}\frac{\partial u_i}{\partial x_j} + \beta T\left(u_j \frac{\partial p}{\partial x_j}\right) \qquad (2\text{-}12)$$

2.2. Simplifying Assumptions

To build precisely our models, it is necessary to take account of simplifying assumptions. For that, we consider that the fluid is Newtonian and incompressible flow. Thus, the stress tensor is proportional to the symmetrical part of the tensor of the rates of deformation. The stress tensor can be reduced to:

$$\overline{\overline{\tau}} = \mu(\overline{\overline{\nabla}}.\vec{V}) \qquad (2\text{-}13)$$

Total conductivity is constant, *i.e.* the heat flow is proportional to the gradient of the temperature:

$$\vec{q} = -\lambda \vec{\nabla} T \qquad (2\text{-}14)$$

The total conductivity λ takes into account the thermal conductivity of the fluid and the turbulent thermal conductivity.

The viscous dissipation flow of the heat is negligible, which results in the inequality according to:

$$\vec{\nabla}.\left(\overline{\overline{\tau}}.\vec{v}\right) \ll |\vec{\nabla}.\vec{q}| \tag{2-15}$$

Boussinesq model can be assumed for the buoyancy force in the momentum equation that the density varies only linearly with the temperature:

$$\rho = \rho_0 \left[1 - \beta \left(T - T_0\right)\right] \tag{2-16}$$

$$\beta = -1/\rho \left(\frac{d\rho}{dT}\right) \tag{2-17}$$

Where β is the thermal expansion coefficient.

Assuming an incompressible fluid

$$\frac{d\rho}{dt} = 0 \tag{2-18}$$

Since the energy source term is not needed in every thermodynamic case, it can also be disregarded: $r = 0$.

2.3. Simplified Equations

Implementing all of the above simplifications, the fundamental equations become:

Continuity Equation:

$$\vec{\nabla}.\rho\vec{V} = 0 \tag{2-19}$$

Momentum Equations:

$$\frac{d\rho\vec{V}}{dt} + (\vec{V}.\vec{\nabla}).\vec{V} = -\frac{1}{\rho_0}.\vec{\nabla}p + \frac{\mu}{\rho_0}\overset{\longrightarrow}{.}(\vec{\nabla}.\vec{V}) + \vec{g}.(1 - \beta\,(T - T_0)) \qquad (2\text{-}20)$$

Energy Conservation Equation:

$$\frac{d\rho CpT}{dt} + \vec{\nabla}.\rho CpT\vec{V} = \vec{\nabla}.(\lambda\vec{\nabla}T) \qquad (2\text{-}21)$$

$$\frac{dT}{dt} + \vec{\nabla}T.\vec{V} = \alpha\,\vec{\nabla}.\vec{\nabla}T \qquad (2\text{-}22)$$

Where:

$$\alpha = \frac{\lambda}{\rho c_p} \qquad (2\text{-}23)$$

These equations do not have an analytical solution; in this manner, for the most part, it utilized rearrangements of these to accomplish a surmised arrangement. The dimensionless number is utilized to complete these disentanglements, as the Reynolds number or the Rayleigh number.

The Reynolds number gives a proportion of the proportion of inertial forces to viscous forces and thus, evaluates the overall significance of these two kinds of forces for given stream conditions:

$$Re = \frac{\rho\,U\,D}{\mu} \qquad (2\text{-}24)$$

The Rayleigh measures the importance of buoyancy-driven flow. When the Rayleigh number is below the critical value for that fluid, the heat transfer is primarily in the form of conduction. When it exceeds the critical value, the heat transfer is primarily in the form of convection.

$$Ra = \frac{g\beta\Delta TC_p\rho^2 L^3}{\mu_k} \qquad (2\text{-}25)$$

For pure natural convection, the Rayleigh number events the strength of the buoyancy-induced flow. When $Ra<10^8$ the condition shows an induced laminar flow and a transition to turbulent flow among $10^8<Ra<10^{10}$.

3. COMPUTATIONAL FLUID DYNAMICS (CFD)

3.1. Need of CFD

Computational Fluid Dynamics (CFD) has become a broadly approved approach for resolving complex fluid flow problems in fluid mechanics and heat transfer but also has become a significant device in chemical and process engineering. In the face of increasing industrial competitiveness and sustainability, the industry is mandatory to decrease the time-to-market, accounting, besides a more energy-efficient, safer, and flexible process design with reduced emissions compared to the actual design. Especially when it approaches novel and enhanced process projects, CFD can support different design possibilities and diminish, for instance, recirculation zones, lowering the final effectiveness of the process. Applying the fundamental laws of mechanics to a fluid offers the governing equations for a fluid. These equations, along with the conservation of energy equation form a set of fixed, nonlinear partial differential equations. It is not possible to resolve these equations analytically for most industrial cases. Conversely, it is possible to get estimated computer-based solutions to the governing equations for a diversity of engineering problems. This is the focus of Computational Fluid Dynamics (CFD). CFD is based on the Navies-Stokes equations. These equations describe how the velocity, pressure, temperature, and density of a moving fluid are linked. Its modeling can help a manager classify hot spots and learn anywhere cold air is being wasted or air is mixing. There are many CFD software, from this we are using ANSYS Fluent 17.0 to modeling the airflow in the solar chimney.

3.2. CFD Strategy

Several commercial CFD programs are available, which included a graphical user interface (GUI) for a user approachable structure of the simulations. Commercial CFD codes are for example CFX, FLUENT, and Flow-works simulation. After the choice of suitable simulation software, the flow problem has to be modeled, beginning by the pre-processing and ending by the post-processing of the numerical simulation. The principal structure of the flow modeling is shown in Fig. (**2.1**) and it can be divided into three main steps, the preprocessing, the solver, and the post-processing.

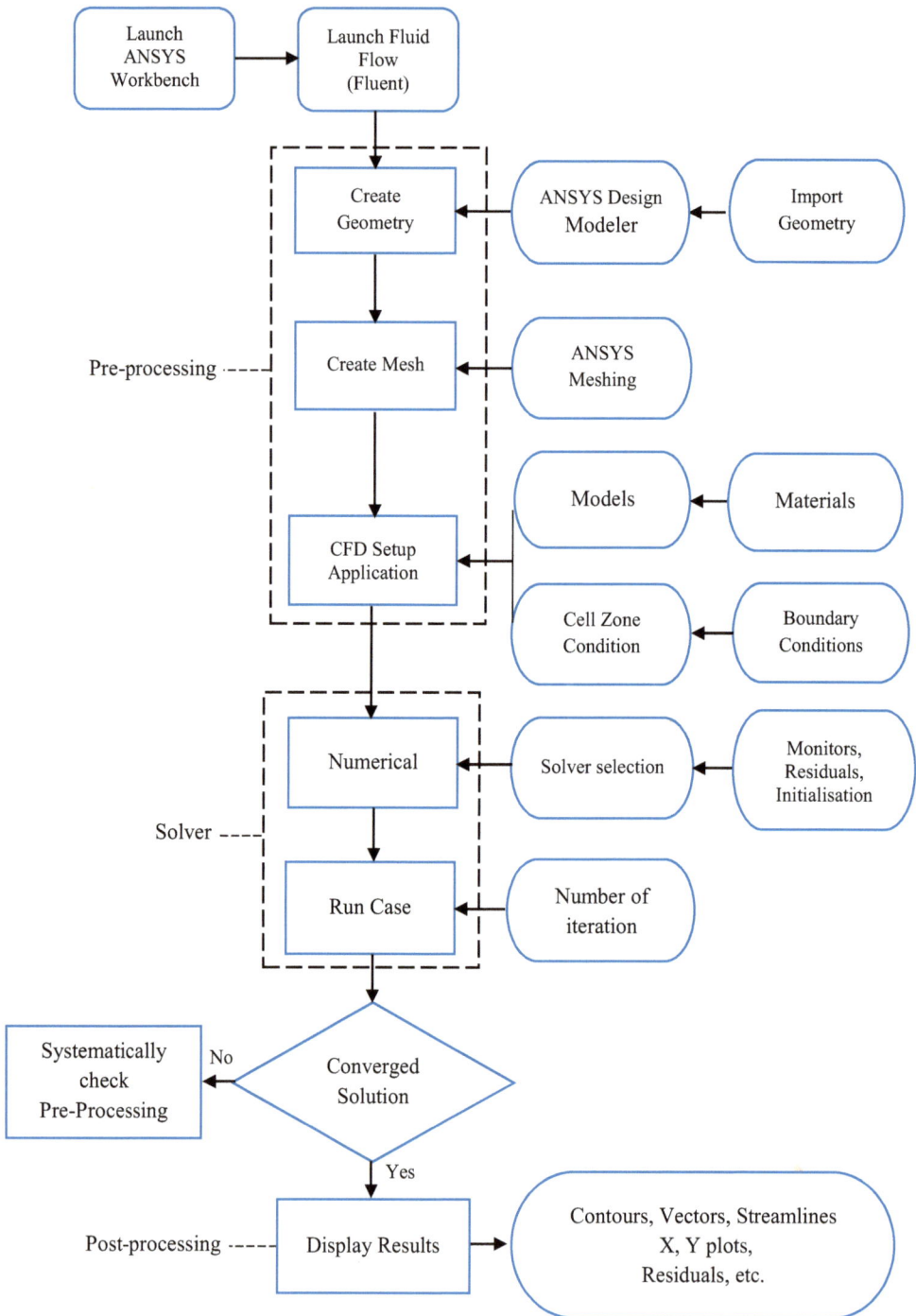

Fig. (2.1). CFD strategy.

3.3. Mesh

A mesh is the spatial discretization of a continuous medium, or as geometric modeling of a field by proportionate and well-defined finite elements. The purpose of a grid is a simplification of a system by a model of this system. It is one of the most important aspects for any CFD simulation, it has two technologies, two dimensional (2D) or three-dimensional mesh (3D). The meshes can be categorized into two groups: structured grids and unstructured meshes (Fig. **2.2**). Programs of numerical calculations based on the finite-difference model require having a structured mesh, while programs based on the finite volume model can use one or the other, although a regular mesh structure is always preferable.

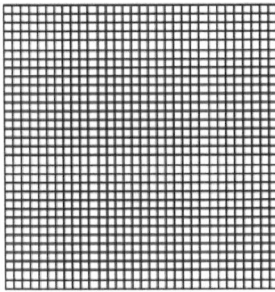

(a) Structured meshing   (b) Unstructured meshing

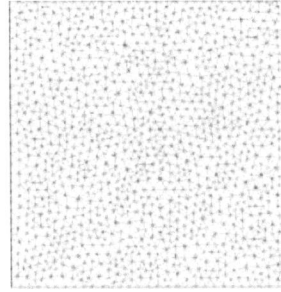

Fig. (2.2). Meshing type.

Unstructured meshes: The main advantage of unstructured grids is that they have the advantage of complying with almost any desired geometry. This facilitates the automation of the mesh generation, which explains the growing popularity of this model. However, unstructured grids require storing more information, and geometric element changes can increase errors in numerical approximations. A popular type of unstructured mesh consists of triangular elements (2D) or tetrahedral (3D). These grids tend to be easier to produce than those composed of quadrilateral elements, but they generally have lower numerical precision.

Structured meshes: When a mesh with a regular structure is used, the main advantage is that the numerical solver is faster than an unstructured mesh. This is because the indexing of points in a regular grid is more direct than in irregular mesh.

There are several general advantages to using a structured grid:

- CFD solvers converge recovered and can give more accurate results when the grid is ranged with the chief flow direction.
- The application of boundary conditions and turbulence models work well when there is a well-defined computational direction normal to a feature such as a wall or awake. Transverse normal are easily defined in a structured grid.

The convergence of the mesh: Discretization introduces errors that depend on the distance Δx between mesh nodes. Accordingly, by increasing the number of nodes, Δx is reduced and so the error. But, it increases the computation time. To determine the number of nodes required to obtain sufficient accurate results, we must increase the resolution of the mesh until the results obtained with a different number of nodes are sufficiently close. We can then consider that we have a grid convergence. To validate our meshes, we tried different resolutions until the results converge.

3.4. Discretization Methods

To numerically solve the differential equations with the partial derivative, established in the preceding section, we will proceed to their discretization to obtain a system of algebraic equations. There are several discretization methods, most known are finite volume method (FVM), Finite element methods (FEM), and Finite difference method (FDM).

The finite volume method (FVM) is a common approach used in FLUENT code. This method was described for the first time by Patankar and Spalding (1971) and was published by Patankar (1980). The principle of this method rests on a technique of discretization, which converts the differential equations with the derivative partial into nonlinear algebraic equations, which can be solved thereafter numerically.

The FVM comprises primarily the following stages:

- The division of the field considered in volumes of control.
- The integral formulation of the differential equations to the derivative partial.
- Writing of the algebraic equations to the nodes of the grid.

3.5. Convergence Criteria

Convergence is determined by the value of residues. These are determined at each iteration for each variable (density, speed, energy) in each elementary mesh cell.

Residue values inform us about the degree of convergence of the equation associated with each variable. There are no absolute criteria for assessing convergence but the value of residues, their evolution, and the value of the unknown are important to make this assessment. The convergence criterion is based on the sum of the normalized residue on the set of points of the computational domain. If the algebraic form of a conservation equation in any control volume momentum equation could be solved exactly, it would be written as:

$$A_p \Phi_p - \sum_i A_i \Phi_i = 0$$

(2-26)

As the solution of each equation in an iterative calculation is based on inaccurate information, originating from initial predicted values and refined through frequent iterations, the right-hand side of the beyond equation is always non-zero. This non-zero value characterizes the error, or residual in the solution of the equation in the control volume:

$$R_p = A_p \Phi_p - \sum_i A_i \Phi_i$$

(2-27)

The total residual is the sum of over all cells in the computational domain of the residuals in each cell:

$$R = \sum_p R_p$$

(2-28)

When R defined in this manner, is reliant on upon the magnitude of the variable presence solved, it is usual to either normalize or scale the total residual to measure its changing value during the solution process.

3.6. Turbulent Models

For turbulent flows, the range of length scales and complexity of phenomena involved in turbulence make most modeling approaches prohibitively expensive; the resolution required to resolve all scales involved in turbulence is beyond what is computationally possible. The primary approach in such cases is to create numerical models to approximate unresolved phenomena. This section lists some commonly used computational models for turbulent flows.

3.6.1. k-ε Model

The k-ε model is a two equations turbulence model, while it just does not achieve well in cases of large adverse pressure gradients. It contains two extra transport equations to represent the turbulent characteristics of the flow. This allows a two-equation model to account for history effects like convection and diffusion of turbulent energy. The first transported variable is the turbulent kinetic energy k. The second transported variable, in this case, is the dissipation rate of the turbulent kinetic energy. It is the variable that determines the scale of the turbulence, whereas the first variable k determines the energy in the turbulence.

3.6.1.1. Standard k-ε Model

The standard k-ε model is a semi-empirical model based on model transport equations for the turbulent kinetic energy k and its dissipation rate ε. The model transport equation for k is derived from the exact equation, while the model transport equation for ε was obtained using physical reasoning and bears little resemblance to its mathematical counterpart. In the derivation of the k-ε model, it was assumed that the flow is fully turbulent, and the effects of molecular viscosity are negligible. The standard k-ε model is therefore valid only for fully turbulent flows. The turbulent kinetic energy, k, and its rate of dissipation, ε, are obtained from the following transport equations:

$$\frac{\partial}{\partial t}(\rho k) + \frac{\partial}{\partial x_i}(\rho k u_i) = \frac{\partial}{\partial x_j}\left[\left(\mu + \frac{\mu_t}{\sigma_k}\right)\frac{\partial k}{\partial x_j}\right] + G_k + G_b - \rho\varepsilon - Y_M + S_k \quad \textbf{(2-29)}$$

And

$$\frac{\partial}{\partial t}(\rho\varepsilon) + \frac{\partial}{\partial x_i}(\rho\varepsilon u_i) = \frac{\partial}{\partial x_j}\left[\left(\mu + \frac{\mu_t}{\sigma_\varepsilon}\right)\frac{\partial\varepsilon}{\partial x_j}\right] + \frac{\varepsilon}{k}C_{1\varepsilon}(G_k + C_{3\varepsilon}G_b) - C_{2\varepsilon}\rho\frac{\varepsilon^2}{k} + S_\varepsilon \quad \textbf{(2-30)}$$

Where:

G_k: generation of turbulence kinetic energy due to the mean velocity gradients, calculated as described in modeling turbulent production in the k-ε model.
G_b: generation of turbulence kinetic energy due to buoyancy, calculated as described in the effects of buoyancy on turbulence in the k-ε model.
Y_M: contribution of the fluctuating dilatation -incompressible turbulence to the

overall dissipation rate, calculated as described in effects of compressibility on turbulence in the k-ε model.

$C_{1\varepsilon}$, $C_{2\varepsilon}$, and 20+5 $C_{3\varepsilon}$ are constants. σ_k and σ_ε are the turbulent Prandtl numbers for k and ε respectively.

S_k and S_ε are user-defined source terms.

The turbulent (or eddy) viscosity μt is computed by combining k and ε as follows:

$$\mu_t = \rho C_\mu \frac{k^2}{\varepsilon}$$ (2-31)

Constants of the Standard k-ε model are presents in Table **2.1**.

Table 2.1. Constants of the standard k-ε model.

$C_{1\varepsilon}$	$C_{2\varepsilon}$	C_μ	σ_k	σ_ε
1.44	1.92	0.09	1.0	1.3

3.6.1.2. Realizable k-ε Model

The realizable k-ε model differs from the standard k- ε in two important ways. The realizable the k-ε model contains an alternative formulation for the turbulent viscosity. A modified transport equation for the dissipation rate, ε, has been derived from an exact equation for the transport of the mean-square vorticity fluctuation. The term "realizable" means that the model satisfies certain mathematical constraints on the Reynolds stresses, consistent with the physics of turbulent flows. Neither the standard k-ε model nor the RNG is a realizable k-ε model. To understand the mathematics behind the realizable k-ε model, consider combining the Boussinesq relationship and the eddy viscosity definition to obtain the following expression for the normal Reynolds stress in an incompressible strained mean flow:

$$\overline{u^2} = \frac{2}{3}k - 2v_t \frac{\partial U}{\partial x}$$ (2-32)

Using equation (2-31) for $v_t \equiv \mu_t/\rho$, we obtain the result that the normal stress, $\overline{u^2}$, which by definition is a positive quantity, becomes negative, that is, "non-realizable", when the strain is large enough to satisfy:

$$\frac{k}{\varepsilon}\frac{\partial U}{\partial x} > \frac{1}{3C_\mu} \approx 3.7 \tag{2-33}$$

Similarly, it can also be shown that the Schwarz inequality for shear stresses

($\overline{u_\alpha u_\beta^2} \le \overline{u_\alpha^2 u_\beta^2}$;   no summation over α and β) can be violated when the mean strain rate is large. The most straightforward way to ensure the realizability (positivity of normal stresses and Schwarz inequality for shear stresses) is to make, C_μ variable by sensitizing it to the mean flow (mean deformation) and the k-ε turbulence model. The notion of variable C_μ is suggested by many modelers including Reynolds and is well substantiated by experimental evidence. For example, C_μ is found to be around 0.09 in the logarithmic layer of equilibrium boundary layers and 0.05 in a strong homogeneous shear flow.

The modeled transport equations for k and ε in the realizable k-ε model are:

$$\frac{\partial}{\partial t}(\rho k) + \frac{\partial}{\partial x_j}(\rho k u_j) = \frac{\partial}{\partial x_j}\left[\left(\mu + \frac{\mu_t}{\sigma_k}\right)\frac{\partial k}{\partial x_j}\right] + G_k + G_b - \rho\varepsilon - Y_M + S_k \tag{2-34}$$

$$\frac{\partial}{\partial t}(\rho\varepsilon) + \frac{\partial}{\partial x_j}(\rho\varepsilon u_j)$$

$$= \frac{\partial}{\partial x_j}\left[\left(\mu + \frac{\mu_t}{\sigma_\varepsilon}\right)\frac{\partial\varepsilon}{\partial x_j}\right] + \frac{\varepsilon}{k}C_{1\varepsilon}C_{3\varepsilon}G_b + \rho C_1 S_\varepsilon - C_2\rho\frac{\varepsilon^2}{k + \sqrt{v\varepsilon}} + S_\varepsilon \tag{2-35}$$

Where:

$$C_1 = \max\left[0.43, \frac{\eta}{\eta + 5}\right] \tag{2-36}$$

$$\eta = S\frac{k}{\varepsilon} \tag{2-37}$$

$$S = \sqrt{2\,S_{ij}S_{ij}} \tag{2-38}$$

As in other k-ε models, the eddy viscosity is computed from:

$$\mu_t = \rho C_\mu \frac{k^2}{\varepsilon} \tag{2-39}$$

The difference between the realizable k-ε model and the standard and RNG k- ε models is that it is no longer constant. It is computed from:

$$C_\mu = \frac{1}{A_0 + A_s \dfrac{kU^*}{\varepsilon}} \tag{2-40}$$

Where:

$$U^* \equiv \sqrt{S_{ij}S_{ij} + \tilde{\Omega}_{ij}\tilde{\Omega}_{ij}} \tag{2-41}$$

And

$$\tilde{\Omega}_{ij} = \Omega_{ij} - 2\,\varepsilon_{ijk}w_k \tag{2-42}$$

$$\Omega_{ij} = \tilde{\Omega}_{ij} - \varepsilon_{ijk}w_k \tag{2-43}$$

Where $\tilde{\Omega}_{ij}$ is the mean rate-of-rotation tensor viewed in a moving reference frame with the angular velocity w_k.

The model constants A_0 and A_s are given by:

$$A_0 = 4.04 \tag{2-44}$$

$$A_s = \sqrt{6}\,\cos\emptyset \tag{2-45}$$

Where:

$$\emptyset = \frac{1}{3}\cos^{-1}(\sqrt{6}W) \tag{2-46}$$

$$W = \frac{S_{ij}S_{jk}S_{ki}}{\tilde{S}^3} \qquad (2\text{-}47)$$

$$\tilde{S} = \sqrt{S_{ij}S_{ij}} \qquad (2\text{-}48)$$

$$S_{ij} = \frac{1}{2}\left(\frac{\partial u_j}{\partial x_i} + \frac{\partial u_i}{\partial x_j}\right) \qquad (2\text{-}49)$$

The constants of the Realizable k-ε model are presented in Table **2.2**.

Table 2.2. Constants of the standard k-ε model.

$C_{1\varepsilon}$	C_2	σ_k	σ_ε
1.44	1.9	1.0	1.2

3.6.1.3. RNG k-ε Model

The RNG k-ε model was derived using a statistical technique called the renormalization group theory. It is similar in form to the standard k-ε model, but includes the following refinements:

- The RNG model has an additional term in its ε equation that improves the accuracy of rapidly strained flows.
- The effect of swirl on turbulence is included in the RNG model, enhancing accuracy for swirling flows.
- The RNG theory provides an analytical formula for turbulent Prandtl numbers, while the standard k-ε model uses user-specified, constant values. While the standard k-ε model is a High-Reynolds number model, the RNG theory provides an analytically derived differential formula for effective viscosity that accounts for Low-Reynolds number effects. Effective use of this feature does, however, depend on the appropriate treatment of the near-wall region. These features make the RNG k-ε model more accurate and reliable for a wider class of flows than the standard k-ε model. The RNG-based k-ε turbulence model is derived from the instantaneous Navier-Stokes equations, using a mathematical technique called "renormalization group" (RNG) methods. The analytical derivation results in a model with constants different from those in the standard k-ε model, and additional terms and functions in the transport equations for k and ε are written as follows:

$$\frac{\partial}{\partial t}(\rho k) + \frac{\partial}{\partial x_i}(\rho k u_i) = \frac{\partial}{\partial x_j}\left(\alpha_k \mu_{eff} \frac{\partial k}{\partial x_j}\right) + G_k + G_b - \rho\varepsilon - Y_M + S_k \qquad \text{(2-50)}$$

$$\frac{\partial}{\partial t}(\rho\varepsilon) + \frac{\partial}{\partial x_i}(\rho\varepsilon u_i) = \frac{\partial}{\partial x_j}\left(\alpha_\varepsilon \mu_{eff} \frac{\partial\varepsilon}{\partial x_j}\right) + \frac{\varepsilon}{k}C_{1\varepsilon}(G_k + C_{3\varepsilon}G_b) - C_{2\varepsilon}\rho\frac{\varepsilon^2}{k} - R_\varepsilon + S_\varepsilon \qquad \text{(2-51)}$$

Where:

G_k: Generation of turbulence kinetic energy due to the mean velocity gradient.

G_b: Generation of turbulence kinetic energy due to buoyancy.

Y_M: Contribution of the fluctuating dilatation incompressible turbulence to the overall dissipation rate.

The quantities α_k and α_ε are the inverse effective Prandtl numbers for k and ε, respectively.

S_k and S_ε are user-defined source terms. The scale elimination procedure in RNG theory results in a differential equation for turbulent viscosity:

$$d\left(\frac{\rho^2 k}{\sqrt{\mu\varepsilon}}\right) = 1.72\frac{\widehat{V}}{\sqrt{\widehat{V}^3 - 1 + C_V}}d\widehat{V} \qquad \text{(2-52)}$$

Where:

$$\widehat{V} = \frac{\mu_{eff}}{\mu} \qquad \text{(2-53)}$$

$$C_V \approx 100 \qquad \text{(2-54)}$$

Equation (2-52) is integrated to obtain an accurate description of how the effective turbulent transport varies with the effective Reynolds number (or eddy scale), allowing the model to better handle Low-Reynolds-number and near-wall flows. In the High-Reynolds-number limit, equation (2-52) gives:

$$\mu_t = \rho C_\mu \frac{k^2}{\varepsilon} \tag{2-55}$$

With $C_\mu = 0.0845$, derived using RNG theory.

It is interesting to note that this value of C_μ is very close to the empirically determined value of 0.09 used in the standard k-ε model. The effective viscosity is computed using the High-Reynolds-number form in equation (2-55). However, there is an option available that allows using the differential relation given in equation (2-53) when it is essential to include Low-Reynolds-number effects.

In general, turbulence is affected by rotation or swirl in the mean flow. The RNG model provides an option to account for the effects of swirl or rotation by modifying the turbulent viscosity appropriately. The modification takes the following functional form:

$$\mu_t = \mu_{t0}\, f(\,\alpha_s, \Omega, \frac{k}{\varepsilon}\,) \tag{2-56}$$

Where μ_{t0} is the value of turbulent viscosity calculated without the swirl modification using either equation (2-53) or equation (2-55).

Ω is a characteristic swirl number evaluated.

α_s is a swirl constant that assumes different values depending on whether the flow is swirl-s dominated or only mildly swirling. This swirl modification always takes effect for axisymmetric, swirling flows, and three-dimensional flows when the RNG model is selected. For mildly swirling flows, α_s is set to 0.07 and cannot be modified. However, for strongly swirling flows, a higher value of α_s can be used.

The inverse effective Prandtl numbers, α_k and α_e, are computed using the following formula derived analytically by the RNG theory:

$$\left|\frac{\alpha - 1.3929}{\alpha_0 - 1.3929}\right|^{0.6321} \left|\frac{\alpha - 2.3929}{\alpha_0 - 2.3929}\right|^{0.3679} = \frac{\mu_{mol}}{\mu_{eff}} \tag{2-57}$$

Where $\alpha_0 = 1.0$.

In the high-Reynolds number limit ($\frac{\mu_{mol}}{\mu_{eff}} \ll 1$):

$\alpha_k = \alpha_\varepsilon \approx 1.393$

The main difference between the RNG and standard k-ε models lies in the additional term in the equation given by:

$$R_\varepsilon = \frac{C_\mu \rho \eta^3 (1 - \frac{\eta}{\eta_0})}{1 + \beta \eta^3} \frac{\varepsilon^2}{k} \tag{2-58}$$

Where:

$$\eta = \frac{Sk}{\varepsilon} \tag{2-59}$$

The effects of this term in the RNG k-ε equation can be seen more clearly by rearranging equation (2-49). Using equation (2-56), the third and fourth terms on the right-hand side of equation (2-49) can be merged, and the resulting equation can be rewritten as:

$$\frac{\partial}{\partial t}(\rho \varepsilon) + \frac{\partial}{\partial x_i}(\rho \varepsilon u_i) = \frac{\partial}{\partial x_j}\left(\alpha_\varepsilon \mu_{eff} \frac{\partial \varepsilon}{\partial x_j}\right) + \frac{\varepsilon}{k} C_{1\varepsilon}(G_k + C_{3\varepsilon} G_b) - C_{2\varepsilon}^* \rho \frac{\varepsilon^2}{k} \tag{2-60}$$

Where $C_{2\varepsilon}^*$ is given by:

$$C_{2\varepsilon}^* = C_{2\varepsilon} + \frac{C_\mu \rho \eta^3 (1 - \frac{\eta}{\eta_0})}{1 + \beta \eta^3} \tag{2-61}$$

The RNG model is more responsive to the effects of rapid strain and streamlines curvature than the standard k-ε model, which explains the superior performance of the RNG model for certain classes of flows. The model constants $C_{1\varepsilon}$ and $C_{2\varepsilon}$ in equation (2-49) have values derived analytically by the RNG theory.

The constants of the RNG-k-ε turbulence model are presents in Table **2.3**.

Table 2.3. Constants of the RNG-k-ε turbulence model.

$C_{1\varepsilon}$	$C_{2\varepsilon}$	C_{μ}	η_0	β
1.42	1.68	0.0854	4.38	0.012

3.6.2. k-kl-ω Transition Model

The k-kl-ω transition model is used to predict boundary layer development and calculate transition onset. This model can be used to effectively address the transition of the boundary layer from a laminar to a turbulent regime.

The k-kl-ω model is considered to be a three-equation eddy-viscosity type, which includes transport equations for turbulent kinetic energy (k_T), laminar kinetic energy (k_L), and the inverse turbulent time scale (ω).

$$\frac{Dk_T}{Dt} = P_{k_T} + R + R_{NAT} - wk_T - D_T + \frac{\partial}{\partial x_j}\left[\left(v + \frac{\alpha_T}{\alpha_k}\right)\frac{\partial k_T}{\partial x_j}\right] \qquad (2\text{-}62)$$

$$\frac{Dk_L}{Dt} = P_{k_L} - R - R_{NAT} - D_L + \frac{\partial}{\partial x_j}\left[v\frac{\partial k_L}{\partial x_j}\right] \qquad (2\text{-}63)$$

$$\frac{Dk_L}{Dt} = C_{\omega 1}\frac{\omega}{k_T}P_{k_T} + \left(\frac{C_{\omega R}}{f_\omega} - 1\right)\frac{\omega}{k_T}(R + R_{NAT}) - C_{\omega 2}\omega^2$$

$$(2\text{-}64)$$

$$+ C_{\omega 3}f_\omega\alpha_T f_\omega^2 \frac{\sqrt{k_T}}{d^3} + \frac{\partial}{\partial x_j}\left[\left(v + \frac{\alpha_T}{\alpha_\omega}\right)\frac{\partial\omega}{\partial x_j}\right]$$

The model constants for the k-kl-ω transition model are listed in Table **2.4**.

Table 2.4. Constant of the k-kl-ω transition model.

$C_{\omega 1}$	$C_{\omega 2}$	$C_{\omega 3}$
0.44	0.92	0.3

3.6.3. Transition SST Model

The transition SST model is based on the coupling of the SST k-ω transport equations with two other transport equations, one for the intermittency and one for the transition onset criteria, in terms of momentum-thickness Reynolds number. An ANSYS, empirical correlation (Langtry and Menter) has been developed to cover standard bypass transition as well as flows in low freestream turbulence environments. In addition, a very powerful option has been included to allow us to enter the own user-defined empirical correlation, which can then be used to control the equation of transition onset momentum thickness Reynolds number.

The transport equation for the intermittency γ is defined as:

$$\frac{\partial(\rho\gamma)}{\partial t} + \frac{\partial(\rho U_{j}\gamma)}{\partial x_j} = P_{\gamma 1} - E_{\gamma 1} + P_{\gamma 2} - E_{\gamma 2} + \frac{\partial}{\partial x_j}\left[\left(\mu + \frac{\mu_t}{\sigma_\gamma}\right)\frac{\partial\gamma}{\partial x_j}\right] \qquad (2\text{-}65)$$

The transition sources are defined as follows:

$$P_{\gamma 1} = C_{a1}F_{l}\rho S|\gamma F_o|^{C_{\gamma 3}} \qquad (2\text{-}66)$$

$$E_{\gamma 1} = C_{e1}P_{\gamma 1}\gamma \qquad (2\text{-}67)$$

Where: S is the strain rate magnitude, F_1 is an empirical correlation that controls the length of the transition region, and C_{a1} and C_{e1} hold the values of 2 and 1, respectively. The destruction/relaminarization sources are defined as follows:

$$P_{\gamma 1} = C_{a1}\rho\Omega_\gamma F_t \qquad (2\text{-}68)$$

$$E_{\gamma 2} = C_{e2}P_{\gamma 2}\gamma \qquad (2\text{-}69)$$

Where Ω is the vorticity magnitude.

The transition onset is controlled by the following functions:

$$Re_V = \frac{\rho y^2 S}{\mu} \qquad (2\text{-}70)$$

$$R_T = \frac{\rho k}{\mu\omega} \qquad (2\text{-}71)$$

$$F_{o1} = \frac{Re_V}{1.193 \, Re_{\theta C}} \tag{2-72}$$

$$F_{o2} = \min(\max(F_{o1}, F_{o1}^4), 2.0) \tag{2-73}$$

$$F_{o3} = \max\left(1 - \left(\frac{R_T}{2.5}\right)^3, 0\right) \tag{2-74}$$

$$F_o = \max(F_{o2} - F_{o3}, 0) \tag{2-75}$$

$$F_t = e^{-\left(\frac{R_T}{4}\right)^4} \tag{2-76}$$

Where y is the wall distance and $Re_{\theta c}$ is the critical Reynolds number where the intermittency first starts to increase in the boundary layer. This occurs upstream of the transition Reynolds number $Re_{\theta t}$ and the difference between the two must be obtained from an empirical correlation. Both the F_1 and $Re_{\theta c}$ correlations are functions of $Re_{\theta t}$.

The constants for the intermittency equation are presented in Table **2.5**.

Table 2.5. Constant of the transition SST model.

C_{a1}	C_{e1}	C_{a2}	C_{e2}	$C_{\gamma3}$	σ_γ
2	1	0.06	50	0.5	1

3.7. Discrete Ordinates (DO) Radiation Model Theory

The discrete ordinates (DO) radiation model solves the radiative transfer equation (RTE) for a finite number of discrete solid angles, each associated with a vector direction $\vec{s}$ fixed in the global Cartesian system (x, y, z). We control the fineness of the angular discretization, analogous to choosing the number of rays for the DTRM. Unlike the DTRM, however, the DO model does not perform ray tracing. Instead, the DO model transforms into a transport equation for radiation intensity in the spatial coordinates (x, y, z). The DO model solves for as many transport equations as there are directions $\vec{s}$. The solution method is identical to that used for the fluid flow and energy equations.

Two implementations of the DO model are available in ANSYS Fluent: uncoupled and (energy) coupled. The uncoupled implementation is sequential in nature and uses a conservative variant of the DO model called the finite-volume scheme and its extension to unstructured meshes. In the uncoupled case, the equations for the energy and radiation intensities are solved one by one, assuming prevailing values for other variables.

The DO model considers the radiative transfer equation (RTE) in the direction $\vec{s}$ as a field equation.

$$\nabla \cdot (I \, (\vec{r},\, \vec{s})\, \vec{s}\,) + (a + \sigma_s) I(\vec{r},\, \vec{s}) = an^2 \frac{\sigma T^4}{\pi} + \frac{\sigma_s}{4\pi} \int_0^{4\pi} I(\vec{r}, \vec{s}\,') \Phi(\vec{s},\vec{s}\,')\, d\Omega' \quad \textbf{(2-77)}$$

ANSYS Fluent also allows the modeling of non-gray radiation using a gray-band model. The RTE for the spectral intensity $I_\lambda(\vec{r},\, \vec{s})$ can be written as:

$$\nabla \cdot (I_\lambda \, (\vec{r},\, \vec{s})\, \vec{s}\,) + (a_\lambda + \sigma_s) I_\lambda(\vec{r},\, \vec{s}) = a_\lambda n^2 I_{b\lambda} + \frac{\sigma_s}{4\pi} \int_0^{4\pi} I_\lambda(\vec{r}, \vec{s}\,') \Phi(\vec{s},\vec{s}\,')\, d\Omega' \quad \textbf{(2-78)}$$

Here λ is the wavelength, a_λ is the spectral absorption coefficient, and $I_{b\lambda}$ is the black body intensity given by the Planck function. The scattering coefficient, the scattering phase function, and the refractive index n are assumed to be independent of the wavelength.

The non-gray DO implementation divides the radiation spectrum into N wavelength bands, which need not be contiguous or equal in extent. The wavelength intervals are supplied and correspond to values in a vacuum (n = 1). The RTE is integrated over each wavelength interval, resulting in transport equations for the quantity $I_\lambda \Delta\lambda$, the radiant energy contained in the wavelength band. The behavior in each band is assumed gray. The black body emission in the wavelength band per unit solid angle is written as follow:

$$[F(0 \to n\lambda_2 T) - F(0 \to n\lambda_1 T)]n^2 \frac{\sigma T^4}{\pi} \quad \textbf{(2-79)}$$

Where $F(0 \to n\lambda T)$ is the fraction of radiant energy emitted by a black body in the wavelength interval from 0 to λ at temperature T in a medium of refractive index. λ_2 and λ_1 are the wavelength boundaries of the band.

The total intensity $I\left(\vec{r},\ \vec{s}\right)$ in each direction $\vec{s}$ at position $\vec{r}$ is computed using

$$I\left(\vec{r},\ \vec{s}\right) = \sum_{k} I_{\lambda_k}\left(\vec{r},\ \vec{s}\right)\Delta\lambda_k \tag{2-80}$$

Where the summation is over the wavelength bands.

Boundary conditions for the non-gray DO model are applied on a band basis. The treatment within a band is the same as that for the gray DO model.

CONCLUSION

In this chapter, we have presented the basic equation for the numerical method in the computational fluid dynamics and we have given some definition for a CFD code and how to choose the appropriate mesh for any structure. Indeed, we have given some definition and formulation for the turbulence models used like Standard k-ε, Realizable k-ε, RNG k-ε, Transition k-kl-ω, and Transition-SST.

CHAPTER 3

Numerical Models Choice and Validation with Anterior Results

1. INTRODUCTION

This chapter reviews the methodology used to model a solar chimney using ANSYS Fluent 17.0 and the validation study to confirm the different models suitable to simulate the airflow through a solar chimney system. A 2D numerical simulation is carried out to analyze and compare the performance of different numerical models and inner details of the solar chimney by taking Kasaeian *et al.* (2014) as a case validation. We use geometric parameters and metrological data similar to Kasaeian *et al.* (2014). To validate our numerical results with his experimental results. The main objective of this chapter is to choose the numerical parameters using the CFD code to investigate the airflow in the solar chimney.

2. DESCRIPTION OF THE PROBLEM

A solar chimney on a small scale was constructed at the University of Tehran, Iran. Tehran city has a geographical length and width of 51.4° and 35.7°, respectively. The geometry based on designed dimensions similar to Kasaeian *et al.* (2014) is mentioned in Fig. (**3.1**). For the pilot plant, the height and diameter of the chimney are 2 m and 20 cm, respectively. The diameter and angle of the collector are 3 m and zero degrees, respectively. In this application, a polycarbonate (PC) pipe has been used with a thickness of 4 mm for the chimney and the glass for the collector. The absorber consisted of 17 pieces of steel with a thickness of 2 mm and 8 mm chipboard wood which were attached together. In the numerical parameters, we will apply only the steel material because it is the material responsible for the heat flux transfer with air. Table **3.1** shows the characteristics of different materials using in this solar chimney. The experimental data of temperature distribution of Kasaeian *et al.* (2014) were recorded on June 19th, 20th, and 21th, 2013 with the same climatic conditions. The ambient temperature was 306 K and the weather conditions were sunny corresponding to global solar radiation equal to 800W.m[-2], with no wind at the same hour

Haythem Nasraoui, Moubarek Bsisa & Zied Driss

(1:30 pm). The conditions of the numerical analysis, including ambient temperature, solar radiation, and air pressure are almost similar to the experimental conditions.

Fig. (3.1). Solar chimney prototype of Kasaeian *et al.* (2014).

Table 3.1. Characteristics of materials.

Material	Density (kg.m^{-3})	Specific Heat (J.kg^{-1}.K^{-1})	Thermal Conductivity (W.m^{-1}.K^{-1})	Absorption Coefficient (m^{-1})	Refractive Index n
Glass	2700	840	0.77	0.87	1.52
Steel	8030	502.48	16.27	1	0
PC	1200	1200	0.2	0.6	1.59
Air	1.22	1006.43	0.024	0	1

3. NUMERICAL MODEL

3.1. Boundary Conditions

A general interpretation was given based on numerical simulation of the solar chimney. A physical model was simulated using ANSYS Fluent 17.0, based on the geometrical dimensions of the solar chimney. The governing equations were solved, assuming symmetric and steady-state conditions. Also, the turbulent model was applied to describe turbulent flow conditions. Furthermore, wall boundary was applied for the chimney with a heat flux of value equal zero to obtain the adiabatic wall and axis boundary was utilized for the axis of the chimney. Wall boundary was used for the absorber and collector and convective heat transfer option was applied for different parts of the device such as collector and absorber. An inlet pressure boundary condition type was specified at the collector inlet. At the chimney exit, a pressure outlet boundary condition was selected. Furthermore, the pressure inlet is equal to p_i=0 Pa and the pressure outlet is equal to p_o=0 Pa. The boundary conditions are illustrated in Fig. (3.2).

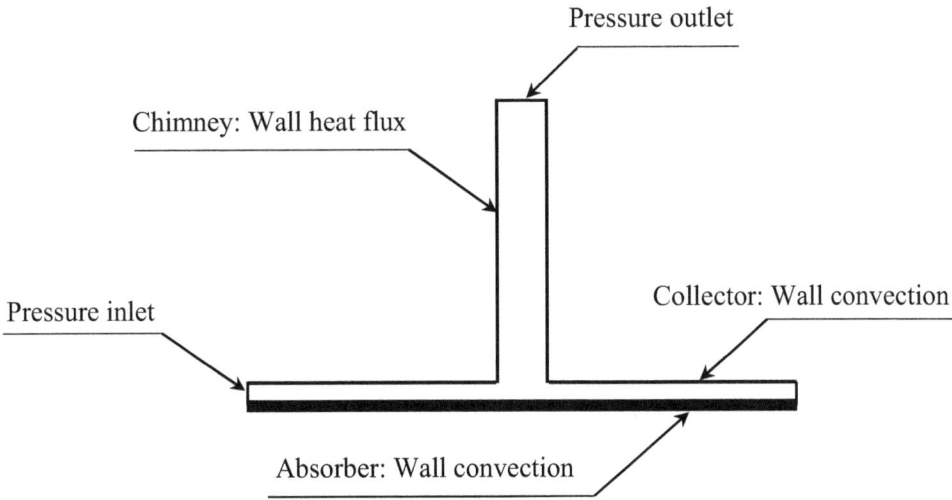

Fig. (3.2). Boundary conditions.

3.2. CFD Parameters

Several parameters govern solar performance and accuracy. The default values are usually good enough to ensure a robust path to a converged simulation. However, occasionally it may be necessary to modify these parameters to achieve better convergence for steady-state simulations via relaxation factors and higher accuracy via differencing schemes. Typically, these parameters are the domain of advanced users. Besides, we take so many iterations to have a good value of residuals (6000 iterations). The convergence criteria are 10^{-5} for all residuals, and configuration does not have problems with the convergence. The used diagram is the scheme "upwind". This scheme takes into account the direction of the flow to determine the size convective on the faces of volumes of controls. The scheme upstream of order one allows certain stability, but it is responsible for numerical diffusion. Therefore, we choose the scheme of the 2nd order consisting of an eccentric scheme upstream, since it minimizes the numerical diffusion. Moreover, the characteristic of the equations makes necessary the use of an algorithm of coupling pressure-speed. The most universal algorithm and more users are undoubtedly the SIMPLE algorithms of Patankar and Spalding (1972). The discrete ordinates method (DO) was used to calculate the radiation heat transfer. DO model has been selected, even it uses most resources and time, and it presents the more realistic obtaining results. Moreover, it is the only model, who allows simulates a semitransparent wall that is going to be important in the next section.

4. MESHING EFFECT

Meshing is one of the most important aspects of any CFD simulation and should be precise enough to consider the complexity in geometry and flow. Regions with high gradients should be identified and the mesh must be refined accordingly. Also, we need to know the available computer memory to handle finer meshes. The quality of the mesh plays a direct role in the quality of the analysis, regardless of the flow solver used. In this section, we are interested in studying the grid dependence of the numerical results in comparison with the experimental results of Kasaeian *et al.* (2014). For this, we have created four types of mesh to choose the better and the appropriate mesh size to use in all the simulations. The meshing parameters are shown in Table **3.2** giving the different parameters for each mesh, nodes number, cell number, elements size, and calculation times. Where T_c is the temperature in the collector at the point position defined with the cylindrical coordinate system defined by r=0.9 m and z=0.03 m. Fig. (**3.3**) shows the views of the different meshing.

Table 3.2. Meshing parameters.

Meshing		First mesh	Second mesh	Third mesh	Fourth mesh
Nodes number		5 103	51 020	238 419	337 567
Cells number		4 663	49 561	229 245	326 105
Elements size (mm)		8	2.4	1.5	1.2
Calculation times (hour)		0.1	3.5	7.5	13
T_c (K)	Numerical	311	313	314	314.5
	Experimental	314			

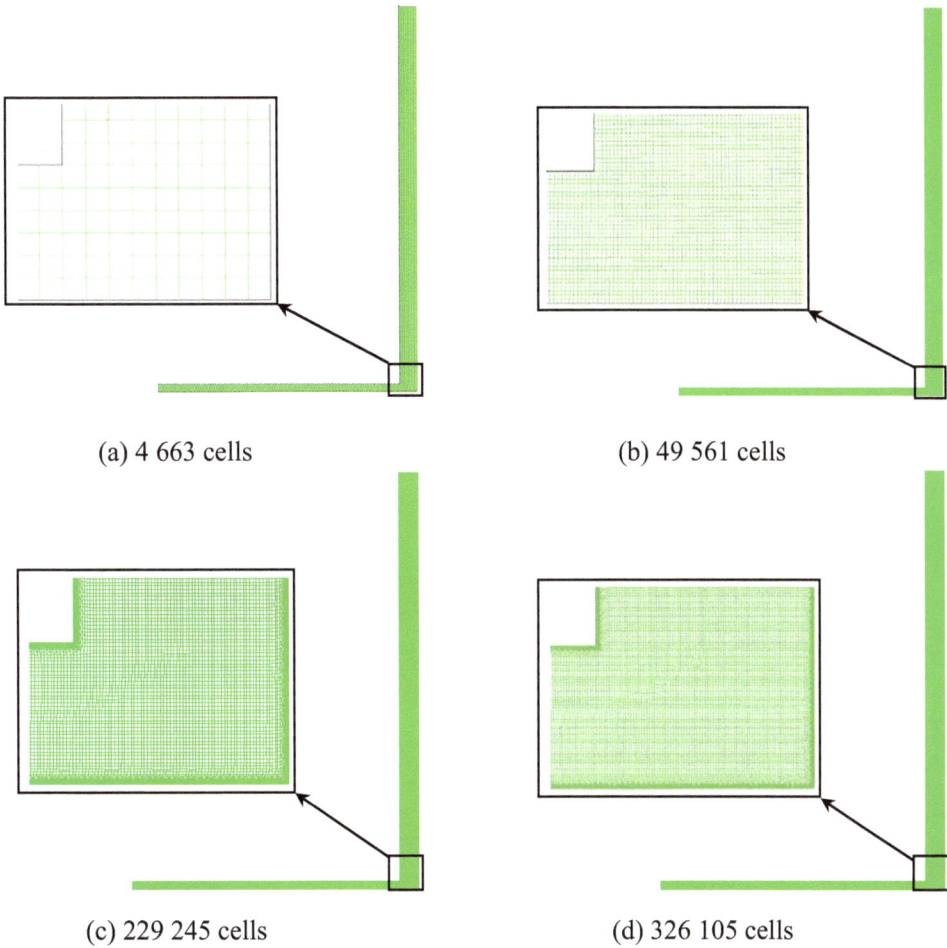

(a) 4 663 cells

(b) 49 561 cells

(c) 229 245 cells

(d) 326 105 cells

Fig. (3.3). Different meshing.

In Fig. (**3.4**), a comparison of the temperature values along the centerline of the collector for the different mesh is illustrated. According to these results, it has been observed that the temperature profile is closer to the experimental results when the mesh becomes finer. Besides, this figure shows a mesh convergence, which is clear with the third and the fourth mesh. In fact, the temperature profiles are confounded, while the simulation data for the fourth mesh is greater. So, we choose the third mesh to minimize the calculation time. From the mesh-size study, it can be concluded that the third mesh is a sufficient spacing to use in modeling the solar chimney. This mesh is also refined on the walls to ensure good resolution in areas of the high gradient of temperature and velocity. Though that numerical simulation is a powerful tool capable of computing fluid flow in a complex configuration, CFD results need to be validated by experimental data.

For this reason, we have compared our predictions with the experimental data of Kasaeian *et al.* (2014).

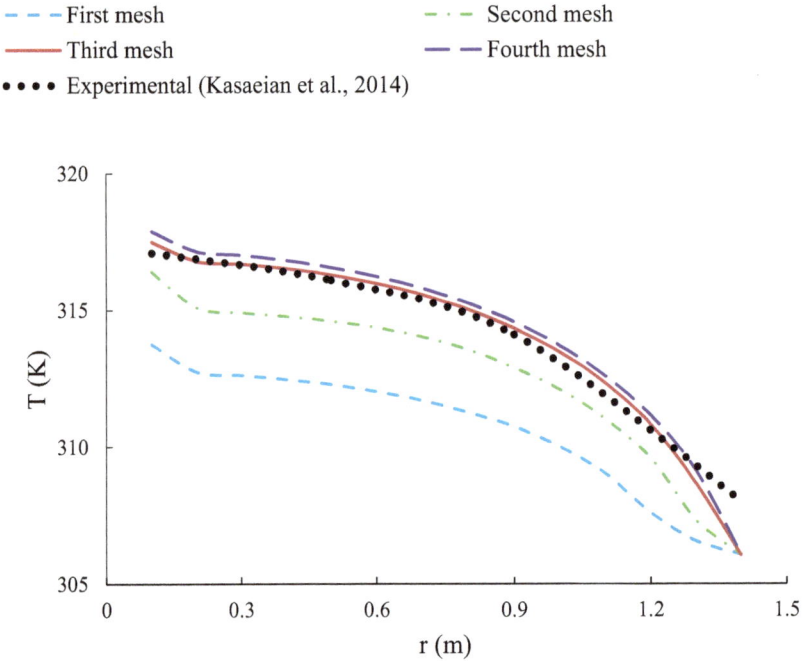

Fig. (3.4). Profiles of the temperature in the collector.

5. TURBULENCE MODEL EFFECT

The solution of the Navies-Stokes equations became unstable with small perturbations. These perturbations develop as the airflow increases, leading to turbulence characterized for quick variation both spatial and temporal. Since these fluctuations can be of small scale and high frequency, they are too computationally expensive to simulate directly in most of the engineering calculations. The choice of the turbulence model is not the same for all kinds of problems. For this reason, we are interested in studying the turbulence model effect in the solar chimney system to understand the capabilities and limitations of the models. This section lists some commonly used computational models. Five models of turbulence have been tested to choose the best model, which can be available to model the solar chimney. Particularly, we have used the standard k-ε model, the realizable k-ε model, the RNG k-ε model, the transition k-kl-ω model, and the transition SST model.

5.1. Temperature

To select the turbulence model, we have compared the temperature of the experimental data with our CFD simulation results. Fig. (**3.5**) shows the temperature profiles of air along the centerline of the collector for the different turbulence model with the experimental data of

Kasaeian *et al.* (2014). According to these results, the temperature profiles of experimental and CFD simulations are similar to two k-ε models. The standard k-ε model is the simplest "complete models" of turbulence. It assumes that the flow is fully turbulent and the molecular viscosity is negligible and the realizable k-ε model is an improvement of the standard k-ε problem. For this, we propose to adopt the realizable k-ε model for further simulations as the best model.

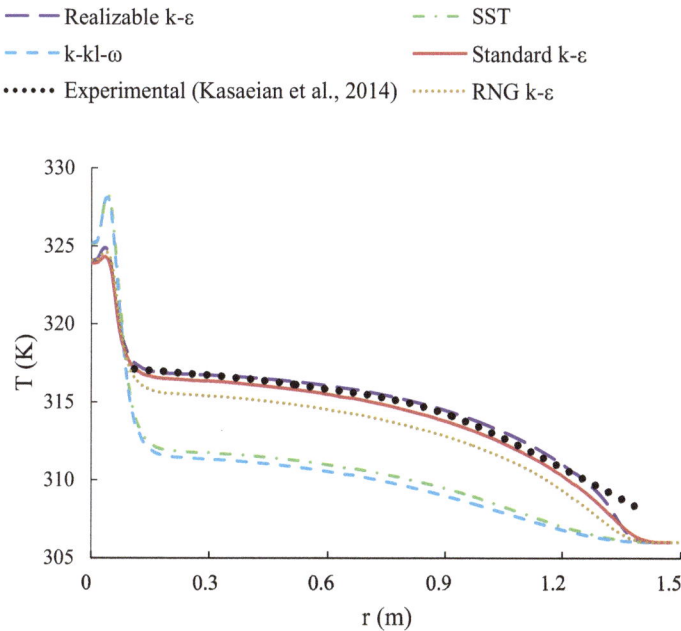

Fig. (**3.5**). Profiles of the temperature in the collector.

The distribution of the temperature in the axisymmetric plane of the solar chimney for different turbulence models is shown in Fig. (**3.6**). According to these results, it has been noted that the change of the turbulence model does not affect the temperature distribution shape. Moreover, for all the cases, the maximum values of temperature are located in the absorber and the chimney axis, otherwise, we observed that the temperature is slightly lower for SST models and k-kl-ω model. The maximum temperature value is equal to T=331 K with the standard k-ε model, to T=331 K with the realizable k-ε model, to T=333 K with the transition-SST model, to T=333 K with the transition-k-kl-ω model and to T=331 K with the RNG k-ε model.

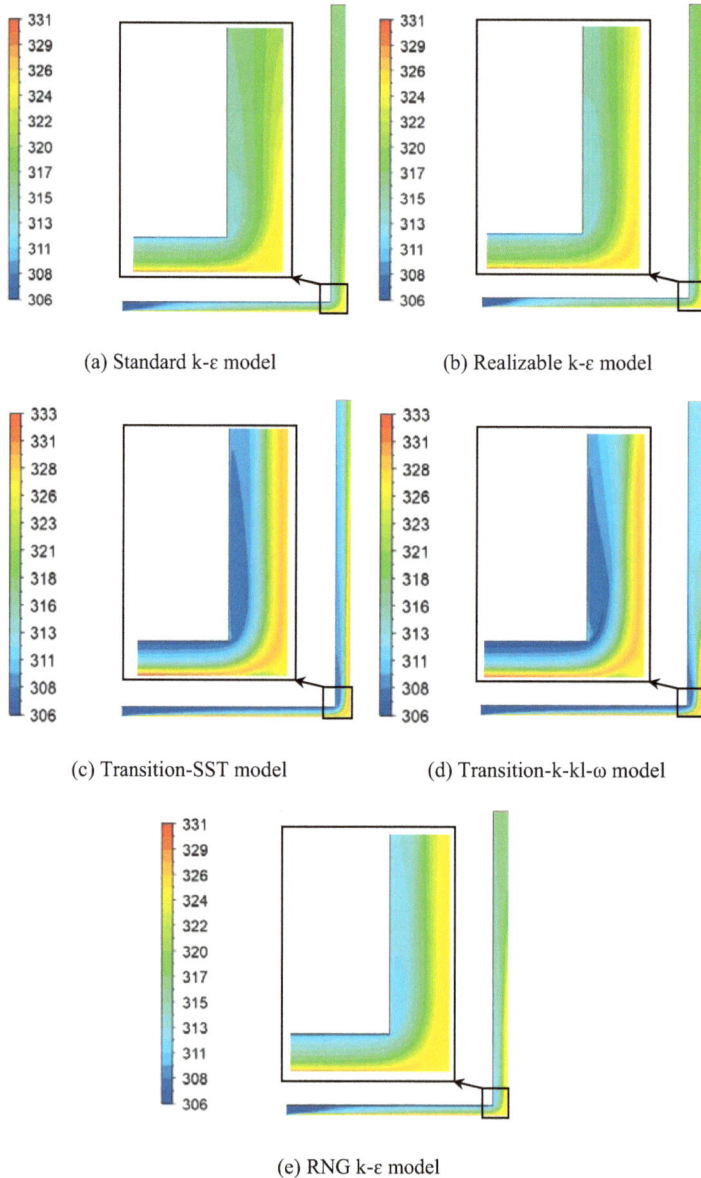

(a) Standard k-ε model

(b) Realizable k-ε model

(c) Transition-SST model

(d) Transition-k-kl-ω model

(e) RNG k-ε model

Fig. (3.6). Distribution of the temperature.

5.2. Magnitude Velocity

The difference between the four turbulence models can be more clearly perceived by comparing the velocity profiles along the chimney axis. As the temperature profiles for the different k-ε models are near the experimental profile, the selection of the right model is a little difficult this is why we use the velocity profiles to specify the adequate turbulence model. Fig. (**3.7**) shows that the magnitude velocity increases by approaching to the chimney base and then remains constant in the chimney portion. These results are the same for all turbulence models except for the k-kl-ω model presenting in the first part of the chimney. Besides, it has been observed that the velocity profile is slightly higher for the realizable k-ε model compared with the other models. For this, we propose to use this model for further work.

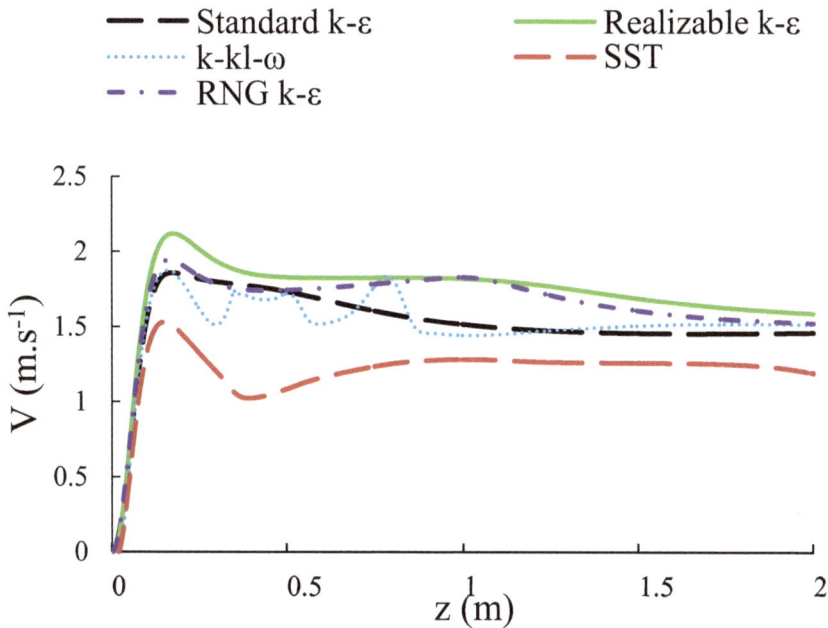

Fig. (3.7). Profiles of the magnitude velocity along the chimney axis.

Fig. (**3.8**) illustrates the velocity fields in the axisymmetric plane of the solar chimney. According to these results, similar distributions are obtained with different turbulence models. The velocity increases in the collector as it approaches the chimney base and then remains constant in the chimney portion. Indeed, it has been observed that the same velocity distribution for all turbulence models where the turbulence zone is located in the base of the chimney where the velocity reaches the maximum value. An acceleration zone located in the base of the chimney has been observed clearly. The increase of the temperature at the chimney base is due to the abrupt velocity change, under the conservation of energy principle. The maximum value of the velocity is equal to V=2.02 m.s^{-1} with the standard k-ε model, to V=2.28 m.s^{-1} with the realizable k-ε model, to V=1.84 m.s^{-1} with the transition-SST model, to V=2.25 m.s^{-1} with the transition--kl- ω model and to V=2.18 m.s^{-1} with the RNG k-ε model.

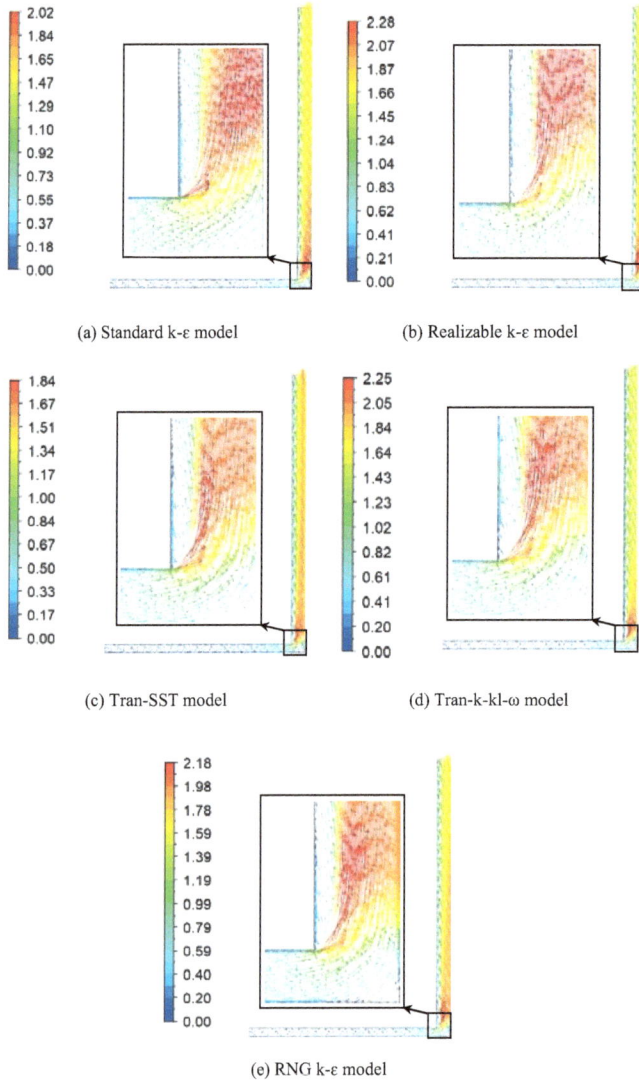

(a) Standard k-ε model   (b) Realizable k-ε model

(c) Tran-SST model   (d) Tran-k-kl-ω model

(e) RNG k-ε model

Fig. (3.8). Velocity fields.

5.3. Static Pressure

The static pressure is defined as the pressure exerted by the air when the bodies on which the pressure was exerted are not in motion. In our system, it is the pressure response to moving the turbine. Fig. (**3.9**) shows the profiles of the static pressure along the centerline of the collector for different turbulence models. It can be seen from these results that the static pressure increases gradually as the air is growing inside the chimney. It also indicates that the minimum static pressure is reached near the chimney base. Likewise, it has been observed that a depression zone, characteristic of the minimum value, appears near the collector where the value of z is low. While in the chimney outlet, a compression zone, characteristic of the maximum values were created, which ensures the outflow of air. From a comparison of these results with the magnitude velocity fields, we see that the static pressure evolves contrary to the magnitude velocity, presenting an acceleration zone, the static pressure is minimal and vice versa.

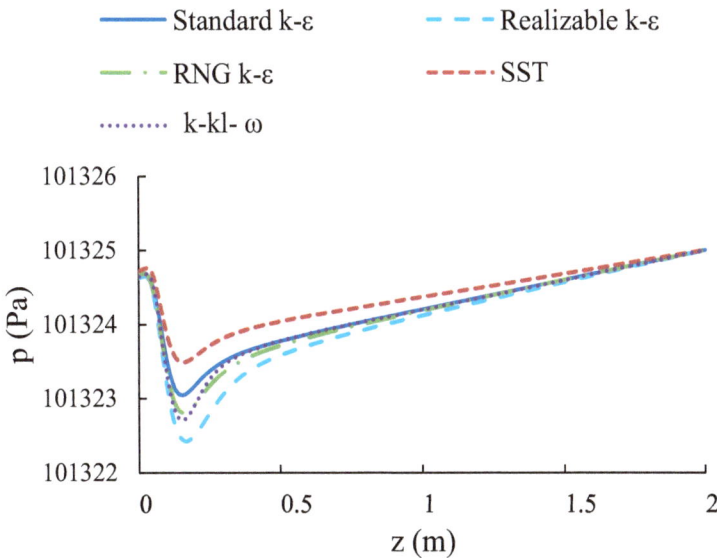

Fig. (3.9). Profiles of the static pressure along the chimney axis.

The distribution of the static pressure in the axisymmetric plane of the solar chimney is illustrated in Fig. (**3.10**). According to these results, a compression zone located in the collector and at the top of the chimney has been observed However, the depression zone is located in the base of the chimney. The static pressure distribution is appreciated to be nominally constant in the collector, decreases considerably at the chimney base, and then starts rising gradually in the chimney portion to meet the hydrostatic pressure value (ambient pressure) at the chimney top. The difference between the five models appears essentially at the depression zone recorded around the base of the chimney where this zone is the largest in the second case with the realizable k-e model. The maximum depression value is equal to p=3.11 Pa with the standard k-ε model, to p= 3.38 Pa with the realizable k-ε model, to p= 2.3 Pa with the transition-SST model, to p=3.17 Pa with the transition-k-kl-ω model and to p=3.24 Pa with the RNG k-ε model.

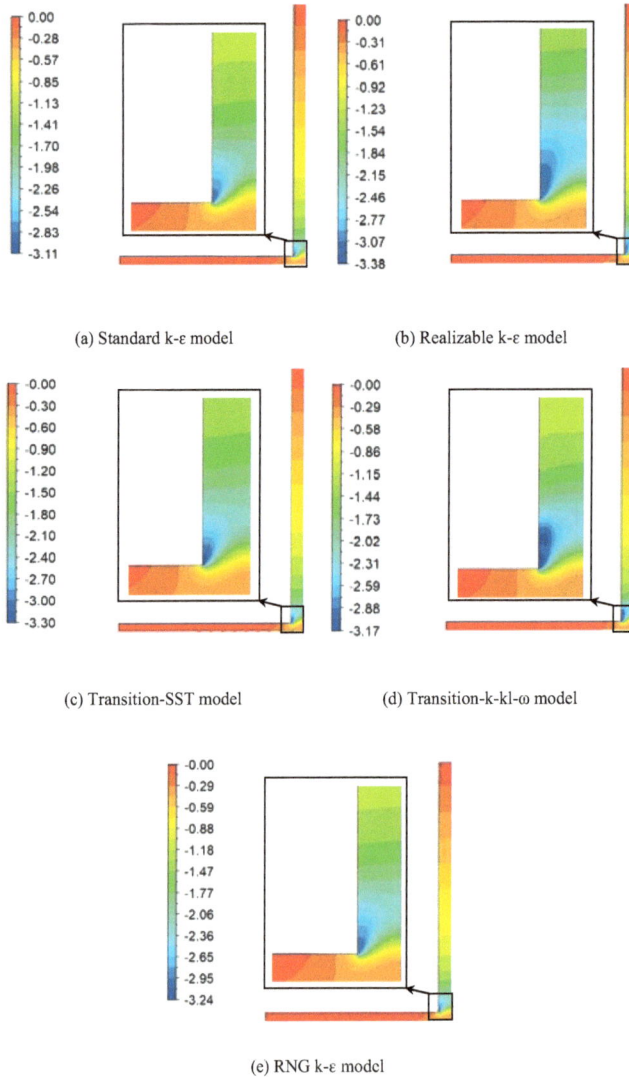

(a) Standard k-ε model   (b) Realizable k-ε model

(c) Transition-SST model   (d) Transition-k-kl-ω model

(e) RNG k-ε model

Fig. (3.10). Distribution of the static pressure.

5.4. Dynamic Pressure

Dynamic pressure is a pressure exerted by the air moving on a surface perpendicular to the direction of air movement across the chimney. Fig. (**3.11**) illustrates the profiles of the dynamic pressure along the chimney axis. According to these results, it has been observed that for the five models, the dynamic pressure evolves in the same way as the magnitude velocity, since the analytic expression of the dynamic pressure is proportional to the square of the magnitude velocity. Besides, it has been noted that in the case with the realizable k-e model, the dynamic pressure is greater along with the chimney. The maximum value of the dynamic pressure appears near to the base of the chimney in z=0.18 m for the five cases. The dynamic pressure is equal to p_d=101327.6 Pa with the realizable k-ε model, but it is equal to p_d= 101326.3 Pa with the transition-SST model.

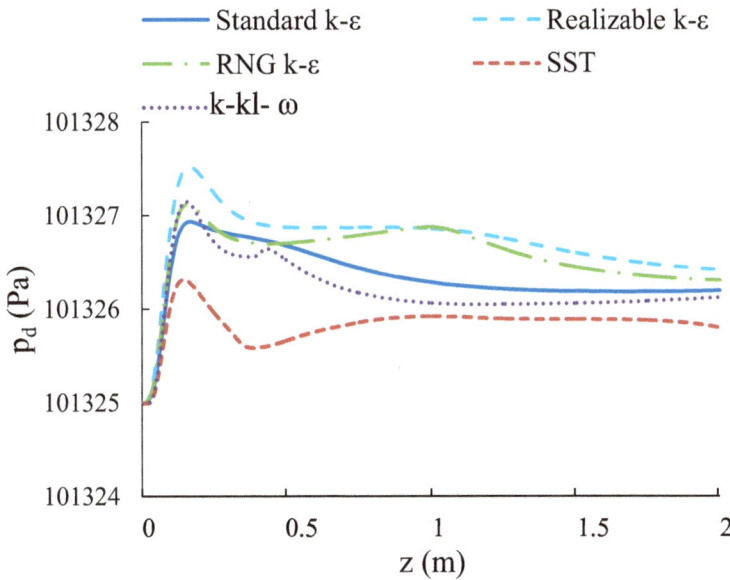

Fig. (3.11). Profiles of the dynamic pressure along the chimney axis.

Fig. (**3.12**) shows the distribution of the dynamic pressure for the five cases of turbulence models. Dynamic pressure is the component of fluid pressure that represents fluid kinetic energy, so it represents the dynamic effects of flow. According to these results, it has been noted that the dynamic pressure has approximately the same behaviors of the magnitude velocity. This fact is due that the two parameters are mathematically linked. Also, it has been observed that the dynamic pressure reaches its maximum value near the chimney base and its value change for the different models. For example, with the transition SST model it is equal to p_d=1.88 Pa, but it is equal to p_d=2.92 Pa for the realizable k-ε model.

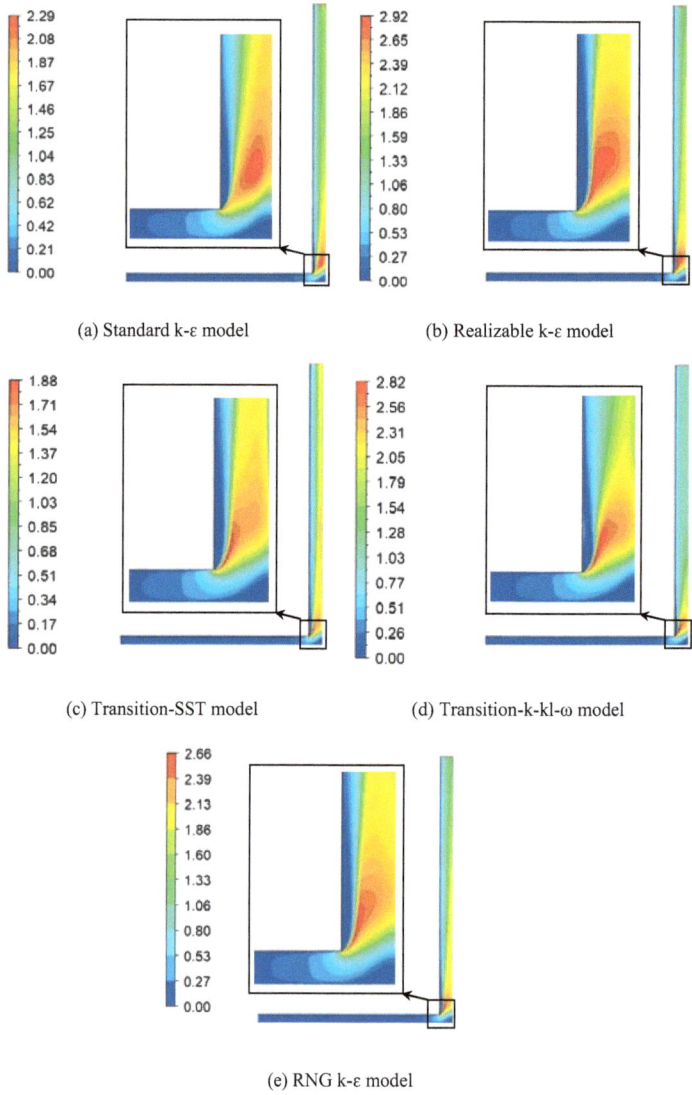

(a) Standard k-ε model   (b) Realizable k-ε model

(c) Transition-SST model   (d) Transition-k-kl-ω model

(e) RNG k-ε model

Fig. (3.12). Distribution of the dynamic pressure.

5.5. Radiation

The radiation is the main parameter for natural convection in the solar chimney. It is the process of emission or energy transmission involving a wave and a particle. Fig. (**3.13**) shows the distribution of the radiation in the axisymmetric plane of the solar chimney system for the five turbulence models. From these results, it is clear that the radiation has the same shape for all models. Along with the collector, the radiation is very greater due to phenomena of the greenhouse effect. Besides, the radiation values are small in the chimney as this latter is isolated. The maximum radiation value is equal to G=3260 W.m^{-2} with the standard k-ε model, to G=3260 W.m^{-2} with the realizable k-ε model, to G=3310 W.m^{-2} with the transition-SST model, to G=3310 W.m^{-2} with the transition-k-kl-ω model and to G=3270 W.m^{-2} with the RNG k-ε model.

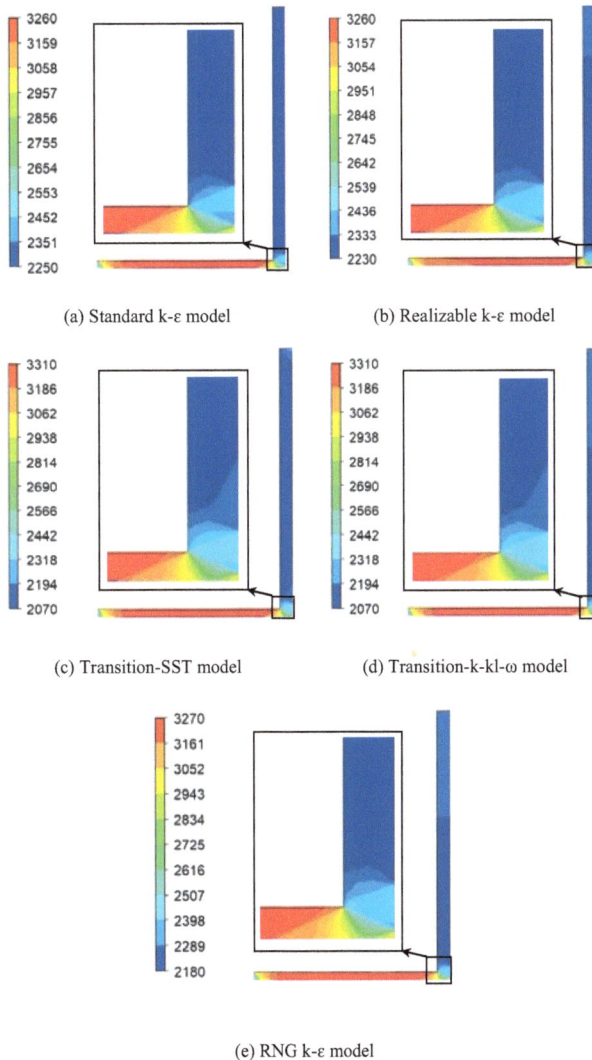

(a) Standard k-ε model   (b) Realizable k-ε model

(c) Transition-SST model   (d) Transition-k-kl-ω model

(e) RNG k-ε model

Fig. (3.13). Distribution of the radiation.

5.6. Turbulent Kinetic Energy

The turbulent kinetic energy is the energy which it possesses due to his motion and due to the turbulence phenomena of the flow. It had been compared to the distribution of this turbulence energy in the solar chimney system, to know what changes have to make in the turbulence models. Fig. (3.14) shows the distribution of the turbulent kinetic energy in the axisymmetric plane of the solar chimney system equipped by the air. From these results, it is clear that the turbulent kinetic energy is found to be very weak in the test section except in the chimney inlet. From a first glance, there may be a small difference between the distributions of the turbulent kinetic energy for each turbulence model. This difference is located in the chimney inlet when the highest value is found in the fourth case with k-kl-ω model. The maximum value of the turbulent kinetic energy is equal to $k=0.271$ $m^2.s^{-2}$ with the standard k-ε model, to $k=0.281$ $m^2.s^{-2}$ with the realizable k-ε model, to $k=0.266$ $m^2.s^{-2}$ with the transition-SST model, to $k=0.411$ $m^2.s^{-2}$ with the transition-k-kl-ω model and to $k=0.285$ $m^2.s^{-2}$ with the RNG k-ε model.

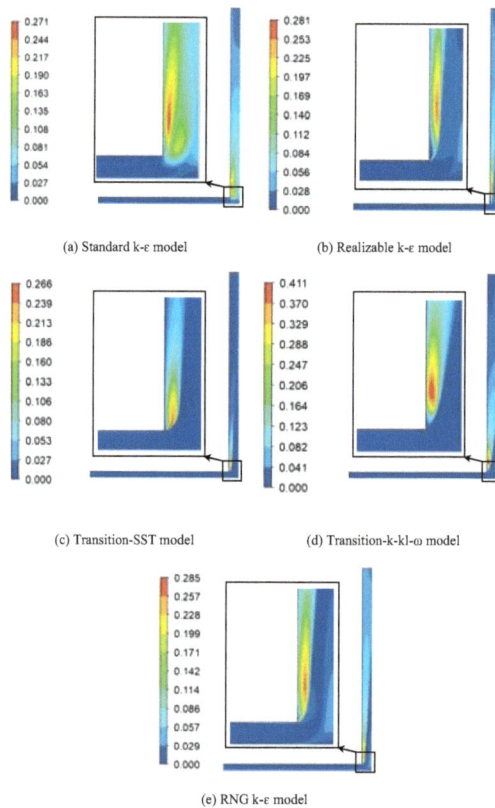

(a) Standard k-ε model   (b) Realizable k-ε model

(c) Transition-SST model   (d) Transition-k-kl-ω model

(e) RNG k-ε model

Fig. (3.14). Distribution of the turbulent kinetic energy.

5.7. Dissipation Rate of the Turbulent Kinetic Energy

The dissipation phenomena are due to the turbulent kinetic energy when it was converted to internal energy for example the thermal energy. Fig. (**3.15**) shows the distribution of the dissipation rate of the turbulent kinetic energy in the axisymmetric plane of the solar chimney. According to these results, it is clear that the variation of the turbulence model has a direct effect on the distribution of the dissipation rate of the turbulent kinetic energy. Indeed, it has been observed that for the two models transition-SST model and transition-k-kl-ω model, the dissipation rate of the turbulent kinetic energy reaches its maximum value at the chimney inlet while for the other models it is negligible. The maximum value of the dissipation rate is equal to $\varepsilon=535$ m^2.s^{-3} with the standard k-ε model, to $\varepsilon=565$ m^2.s^{-3} with the realizable k-ε model, to $\varepsilon=17.9$ m^2.s^{-3} with the transition-SST model, to $\varepsilon=0.967$ m^2.s^{-3} with the transition-k-kl-ω model and to $\varepsilon=551$ m^2.s^{-3} with the RNG k-ε model.

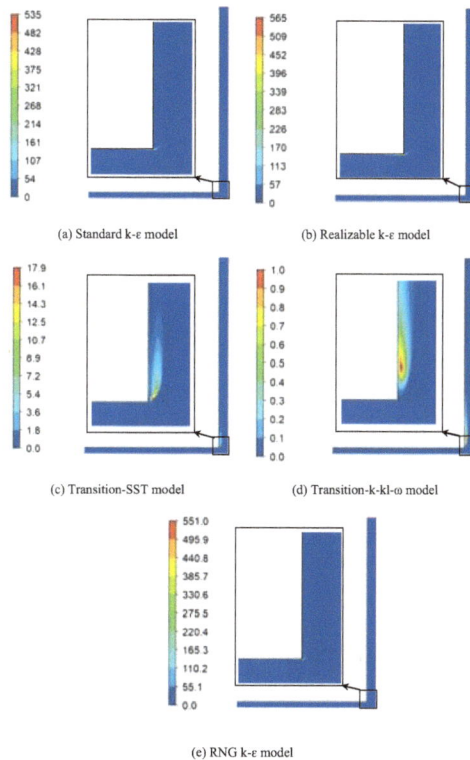

(a) Standard k-ε model

(b) Realizable k-ε model

(c) Transition-SST model

(d) Transition-k-kl-ω model

(e) RNG k-ε model

Fig. (3.15). Distribution of the dissipation rate of the turbulent kinetic energy.

6. HEAT TRANSFER MODE EFFECT IN THE ABSORBER

The absorber is an important component of the solar chimney. It is responsible for the air heat in the collector. Thus, it is responsible for the greenhouse effect which is the base of the principle of solar chimney. In this section, we are interested in the study of the effect of the heat transfer model in the absorber on the performance of the airflow in the solar chimney. The heat transfer model is studied in this part for different modes found in ANSYS Fluent 17.0 consisting of the convection mode, the radiation mode, the mixed-mode, and the heat flux mode. The results of these models are compared with the anterior experimental result. In all the models, ambient pressure and temperature of T=306 K is specified at the inlet and the outlet. These boundary conditions are chosen similar to the experimental conditions of Kasaeian *et al.* (2014). The DO radiation model is incorporated into all of the computer simulations.

6.1. Temperature

Fig. (**3.16**) shows the profiles of the temperature along the centerline of the collector for different heat transfer mode. According to these results, it has been noted that the temperature remains the same shape of profile for all heat transfer modes. The temperature has the minimum value in the collector inlet and it is equal to the ambient temperature T=306K. Then, it increases gradually until reaching its maximum value in the collector outlet. The difference between the four models appears in the maximum value. There are extremely high-temperature levels for the case with constant heat flux. But, the temperature values for the other three cases are much more realistic to the operation of a solar chimney. Furthermore, the heat transfer by convection mode presents results nearest to the experiment data. For this reason, we propose to choose this mode for the further simulations.

The distribution of the temperature in the axisymmetric plane of the solar chimney for different turbulence model is shown in Fig. (**3.17**). These results confirm the results presented in Fig. (**3.16**) and showing the gradual increase of the temperature along with the collector. Indeed, it has been noted that the temperature is highest in the absorber and the axis of the chimney, especially in the lower part of the chimney. The variation of the transfer mode does not affect the general shape of the temperature distribution but the effect is shown in the value where the highest value appears with the radiation mode. The maximum value of the temperature is equal to T=349 K with the heat flux mode, to T=331 K with the convection mode, to T=352 K with the radiation mode, and T=324 K with the mixed mode.

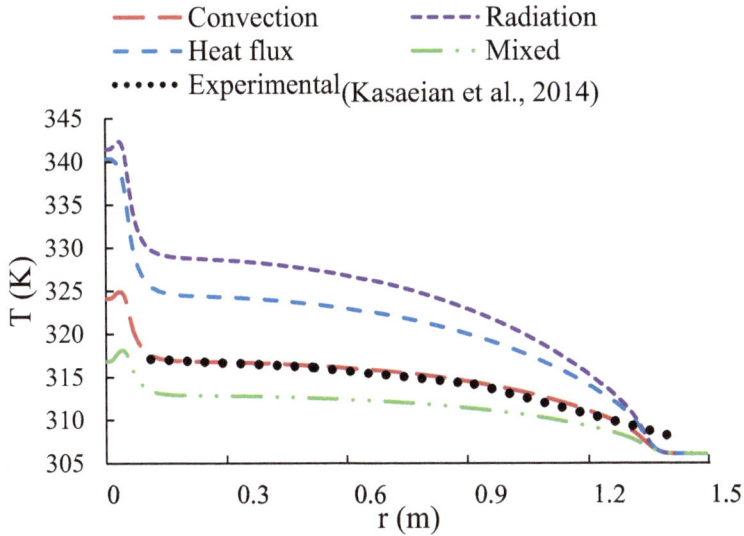

Fig. (3.16). Profiles of the temperature in the collector.

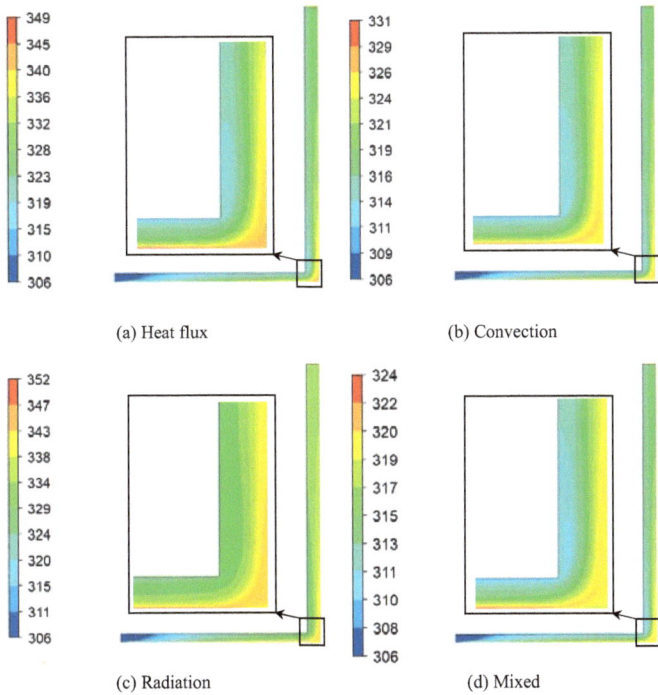

(a) Heat flux   (b) Convection

(c) Radiation   (d) Mixed

Fig. (3.17). Distribution of the temperature.

6.2. Magnitude Velocity

Fig. (**3.18**) illustrates the profiles of the magnitude velocity in the centerline of the collector for the different heat transfer modes. From these results, it has been observed that the magnitude velocity is found weak in the inlet of the collector and increases gradually along with the collector. A flow acceleration appears in the collector outlet near the chimney base. The difference between the four modes appears in the maximum value of the magnitude velocity in the collector outlet. The greatest value is obtained with the radiation model. However, the lowest value is recorded with a mixed model.

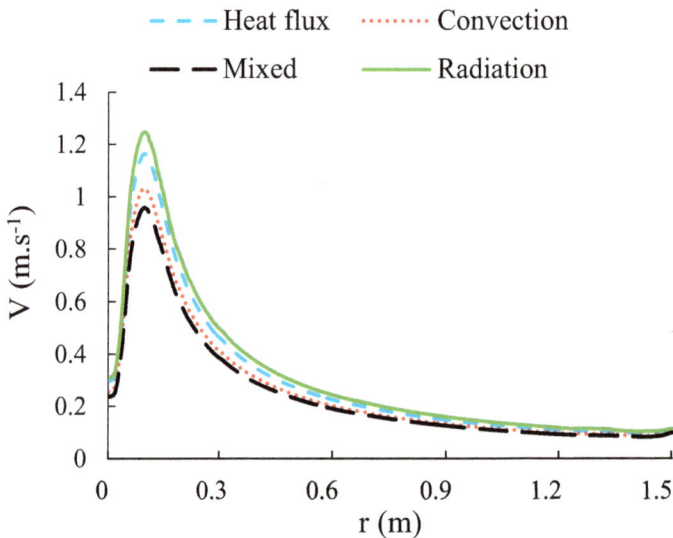

Fig. (3.18). Profiles of the magnitude velocity in the collector.

The distribution of the magnitude velocity in the axisymmetric plane of the solar chimney for the different heat transfer modes is shown in Fig. (**3.19**). According to these results, a similar distribution in the shape has been observed for all cases. The magnitude velocity is low in the system inlet and increases progressively in the collector until reaching the maximum value in the collector outlet corresponding to the chimney inlet. In this area, an acceleration zone has been observed and it remains constant along with the chimney. Also, it has been noted that the difference between the four models appears in the maximum value of the velocity where the maximum value has been obtained $V=2.58$ m.s^{-1} in the case with heat flux. However, with a mixed model it reaches $V=2.11$ m.s^{-1}.

(a) Heat flux

(b) Convection

(c) Radiation

(d) Mixed

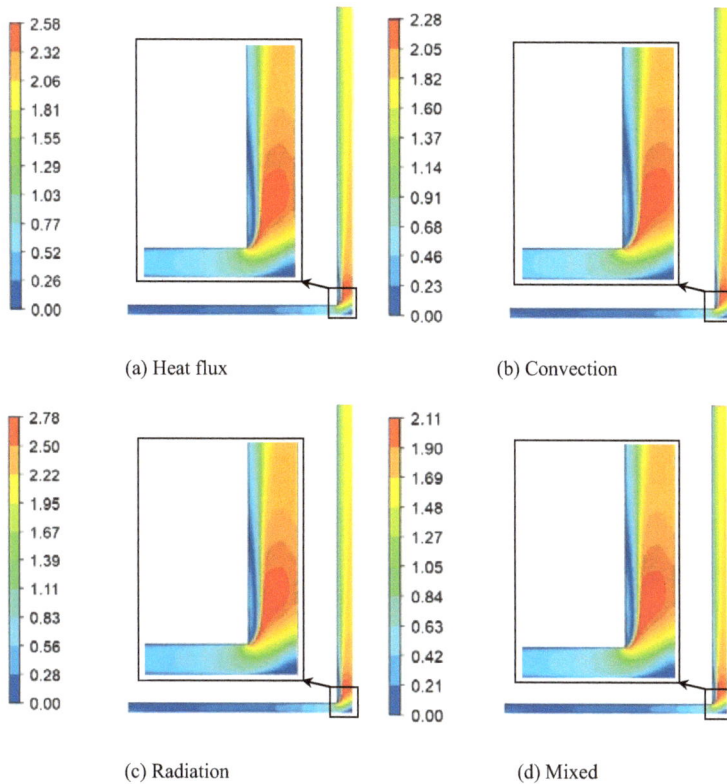

Fig. (3.19). Distribution of the magnitude velocity.

6.3. Radiation

Fig. (**3.20**) shows the distribution of the radiation in the axisymmetric plane of the solar chimney system for the four heat transfer modes applied in the absorber. From these results, it is clear that the radiation has the same shape for all models. Along with the collector, the radiation is very important due to the phenomena of the greenhouse effect. Besides, the radiation is null in the chimney since this latter is isolated. In these conditions, the maximum radiation value is equal to $G=3620$ W.m^{-2} with the heat flux mode, to $G=3260$ W.m^{-2} with the convection mode, to $G=4450$ W.m^{-2} with the radiation mode and to $G=3110$ W.m^{-2} with the mixed mode. The amount of energy absorbed by the absorber and the emissivity of the absorber material is only taken into account when the DO radiation model is applied.

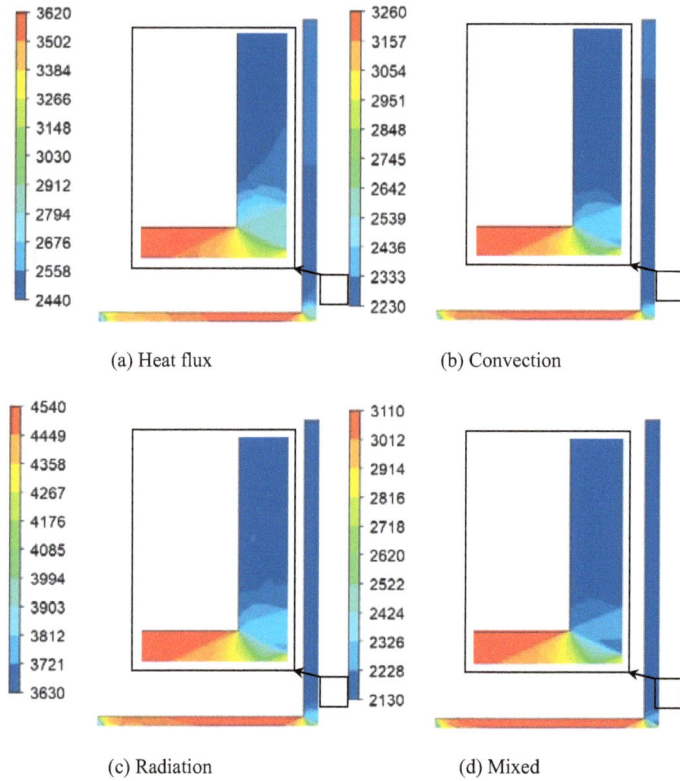

(a) Heat flux   (b) Convection

(c) Radiation   (d) Mixed

Fig. (3.20). Distribution of the radiation.

6.4. Enthalpy

Enthalpy function corresponds to the total energy of the thermodynamic system. It includes the internal energy as the energy needed to create the system, which is added the work that this system must take against the external pressure to fill its volume. Fig. (**3.21**) shows the distribution of the enthalpy in the axisymmetric plane of the solar chimney. According to these results, it is clear that the enthalpy is similar to the temperature for all heat transfer modes. Particularly, it presents the highest values in the absorber and the axis of the chimney. Then, it decreases near the chimney walls due to the heat loss outside the system. The maximum enthalpy value is equal to h=49700 J.kg^{-1} with the heat flux mode, to h=32400 J.kg^{-1} with the convection mode, to h=53500 J.kg^{-1} with the radiation mode and h=24000 J.kg^{-1} with the mixed mode.

(a) Heat flux

(b) Convection

(c) Radiation

(d) Mixed

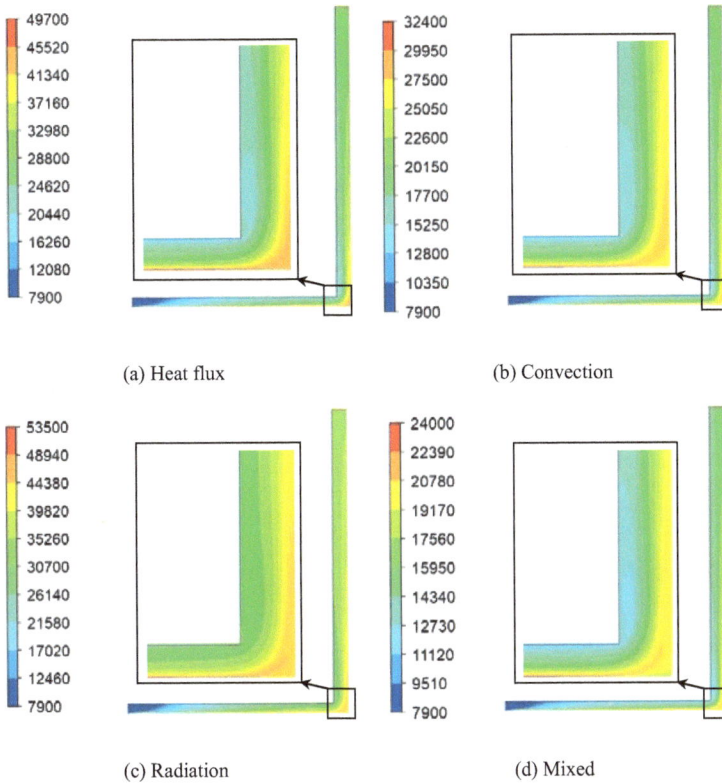

Fig. (3.21). Distribution of the enthalpy.

6.5. Static Pressure

The distribution of the air static pressure is informative about the nature of the flow. Fig. (**3.22**) illustrates the distribution of the static pressure in the axisymmetric plane of the solar chimney system for different heat transfer modes at the absorber. According to these results, the static pressure is the most informative in the area in the collector. Once the air enters the chimney, the static pressure increases as the flow exit the system. Also, it has been observed that the compression zone is located in the collector inlet. However, the depression zone is located in the corners of the chimney inlet. The difference between the four models appears in the minimum value of the static pressure in the base of the chimney. The static pressure value is equal to p= -4.31Pa in the first case with the heat flux mode. However, it is equal to p= -2.9Pa with the fourth case with the mixed mode.

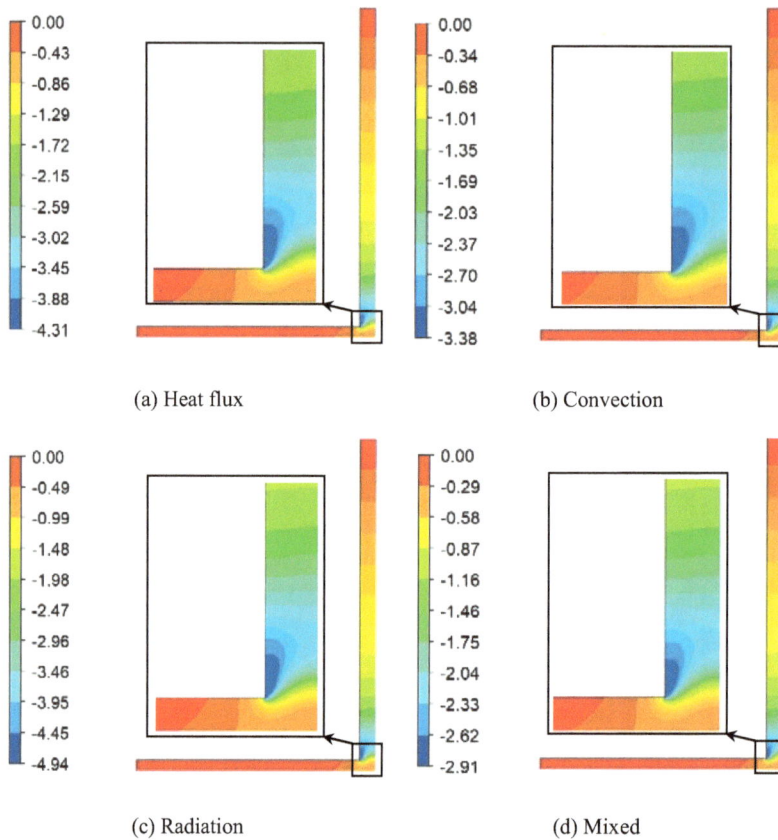

(a) Heat flux

(b) Convection

(c) Radiation

(d) Mixed

Fig. (3.22). Distribution of the static pressure.

6.6. Dynamic Pressure

Fig. (**3.23**) shows the distribution of the dynamic pressure in the axisymmetric plane of the solar chimney for different heat transfer mode in the absorber. According to these results, it has been noted that the dynamic pressure has approximately the same behaviors of the magnitude velocity. This fact is due that these two parameters are mathematically linked. Globally, it has been observed that the dynamic pressure increases gradually in the collector until reaching its maximum value near to the chimney base. Indeed, it is clear that the dynamic pressure presents the same distribution for all heat transfer modes. The difference is founded at the maximum value, which is equal to pd=3.73 Pa with the heat flux mode, to pd=2.92 Pa with the convection mode, to pd=4.33 Pa with the radiation mode and to pd=2.51 Pa with the mixed mode.

(a) Heat flux

(b) Convection

(c) Radiation

(d) Mixed

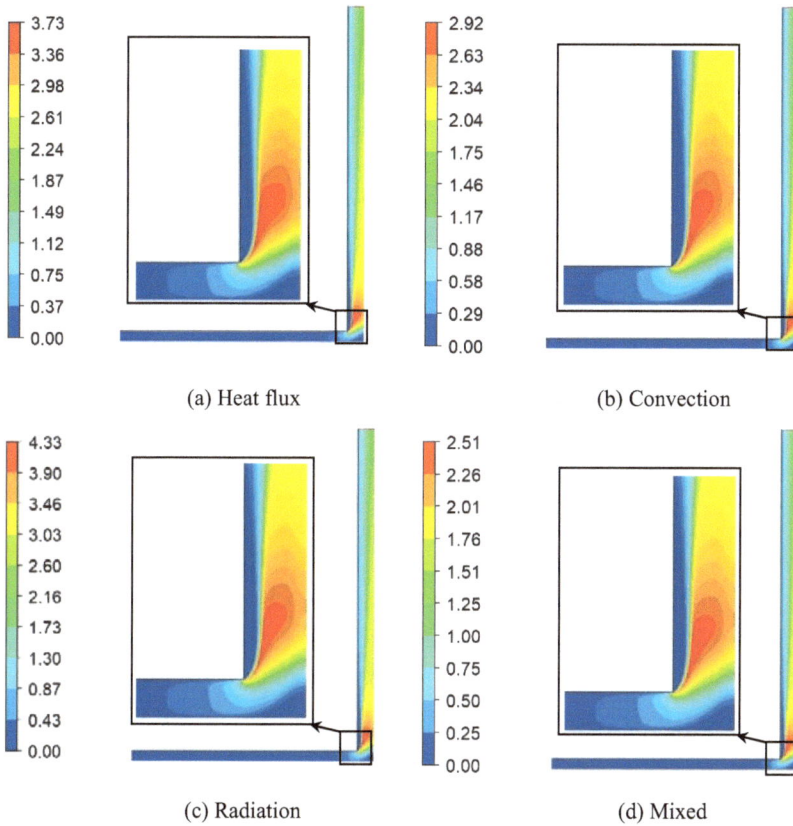

Fig. (3.23). Distribution of the dynamic pressure.

6.7. Turbulent Kinetic Energy

Fig. **(3.24)** shows the distribution of the turbulent kinetic energy in the axisymmetric plane the solar chimney for the different heat transfer modes. According to these results, it has been observed that the wake characteristics of the maximum value of the turbulent kinetic energy appear near the chimney base wall. Elsewhere, the turbulent kinetic energy reduces to a lower value and eventually to zero at the collector wall. The increase of the turbulent kinetic energy near the chimney base wall may be due to the high turbulence production near the chimney base wall as a result of strong shear in the thermal boundary layer. Indeed, it is interesting to note that the turbulent kinetic energy changes at the chimney base, with the change of the heat transfer mode. The highest value appears in the case with radiation mode, but the lowest value appears in the mixed mode. Indeed, it is important to note that the change of the turbulent kinetic energy due to the heat transfer mode in the absorber is the same are shown in the distribution of the magnitude velocity. Fig. (**3.24**). Distribution of the turbulent kinetic energy.

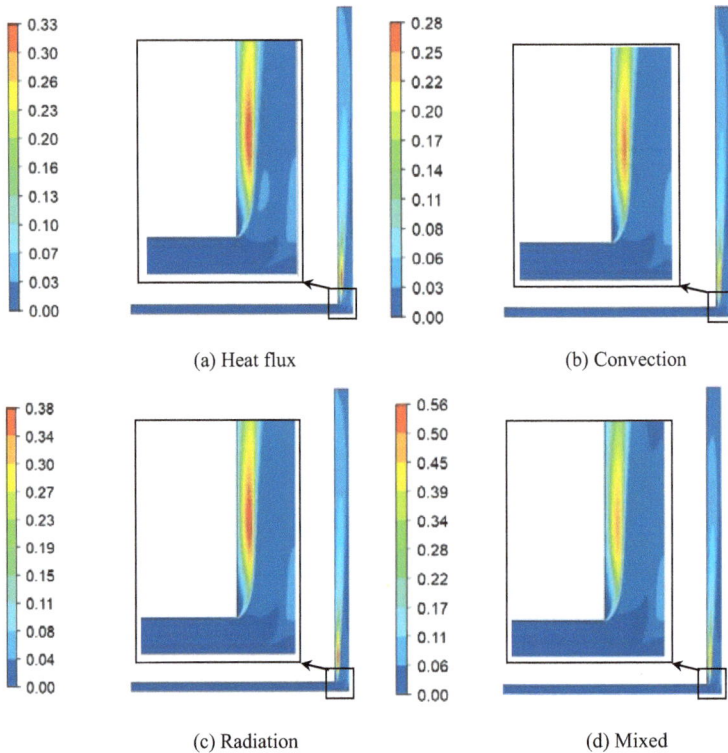

(a) Heat flux

(b) Convection

(c) Radiation

(d) Mixed

Fig. (3.24). Distribution of the turbulent kinetic energy.

6.8. Dissipation Rate of the Turbulent Kinetic Energy

Fig. (**3.25**) shows the distribution of the dissipation rate of the turbulent kinetic energy in the axisymmetric plane of the solar chimney for the different heat transfer modes in the absorber. The dissipation phenomena are due to the turbulent kinetic energy when it was converted to internal energy for example the thermal energy. This fact is due to the important value of the viscous shear stress that appeared on the airflow in the solar chimney. According to these results, it is clear that the dissipation rate of the turbulent kinetic energy presents a low value in all cases. Since the change of the heat transfer mode has not a great effect on the turbulent kinetic energy distribution, it has been noted that the dissipation rate of the turbulent kinetic energy presents the same distribution for all heat transfer modes. The difference is illustrated essentially at the maximum value which is equal to $\varepsilon=698$ m^2.s^{-3} with the heat flux mode, to $\varepsilon=565$ m^2.s^{-3} with the convection mode, to $\varepsilon=786$ m^2.s^{-3} with the radiation mode and $\varepsilon=496$ m^2.s^{-3} with the mixed mode.

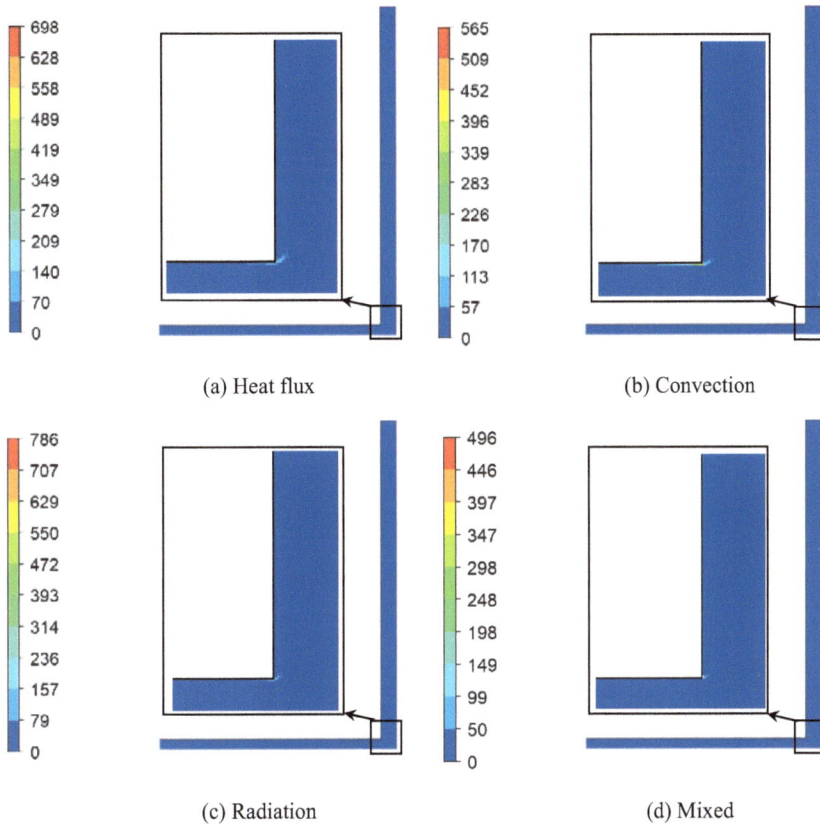

(a) Heat flux   (b) Convection

(c) Radiation   (d) Mixed

Fig. (3.25). Distribution of the dissipation rate of the turbulent kinetic energy.

7. HEAT TRANSFER MODE EFFECT IN THE COLLECTOR

The heat transfer in the collector is an important factor to determine the efficiency of the collector and the efficiency of the solar chimney. The heat transfer mode is studied in this part for the four modes studied in the last part which concern the convection, the radiation, the mixed, and the heat flux. The results of these models are compared with the anterior experimental results. In all the models, ambient pressure and temperature of T=306 K is specified at the inlet and outlet. These boundary conditions are chosen by considering to experimental application of Kasaeian *et al.* (2014). The DO radiation model has incorporated into all of the simulations the temperature and it is compared for the different models.

7.1. Temperature

Fig. (**3.26**) presents the profiles of the temperature along the centerline of the collector for different heat transfer modes. According to these results, it is clear that the change of the heat transfer mode does not affect the evolution of the temperature profile. However, the difference between the four modes is shows in the maximum value. the maximum value is localized near the base of the chimney where the radial position is equal to r=0 m. Then, it decreases until the minimal value recording in the collector inlet. The minimum value of the temperature is equal to the ambient temperature T=306 K. The extremely high temperature is recorded in the case with radiation and mixed model, but the temperature values for the two other cases are much more realistic to the solar chimney operation.

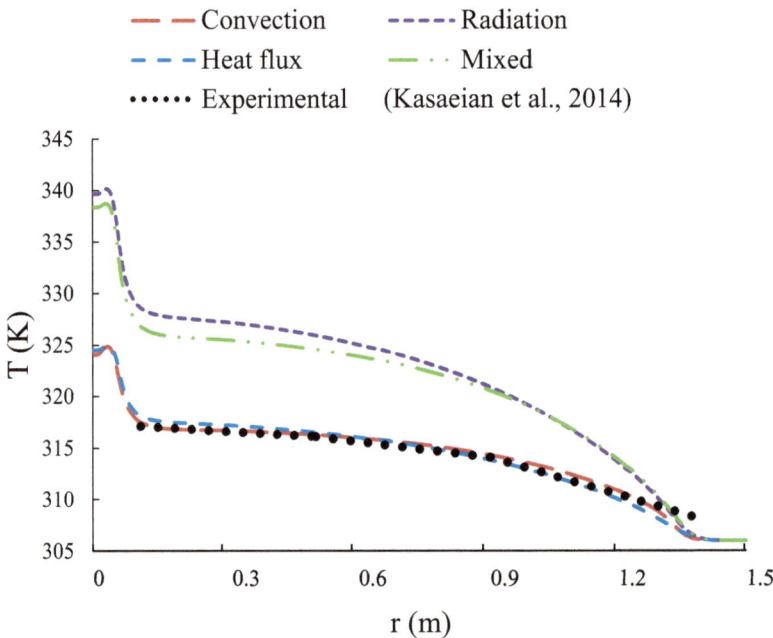

Fig. (3.26). Fields of the temperature in the collector.

The distribution of the temperature in the axisymmetric plane of the solar chimney for different heat transfer mode in the collector is shown in Fig. (**3.27**). According to these results, it has been noted that for the four heat transfer modes, the shape of the temperature distribution is not very affected. The temperature is equal to the ambient in the collector inlet and increases gradually along with the collector until reaching its maximum value in the chimney base. Also, it has been observed that the temperature near the absorber is bigger than that near to the collector. Besides, the difference between the four cases appears in the maximum value of temperature. In fact, the maximum value of the temperature is equal to T=331 K with the heat flux mode, to T=331 K with the convection mode, to T=351 K with the radiation mode, and T=350 K with the mixed mode.

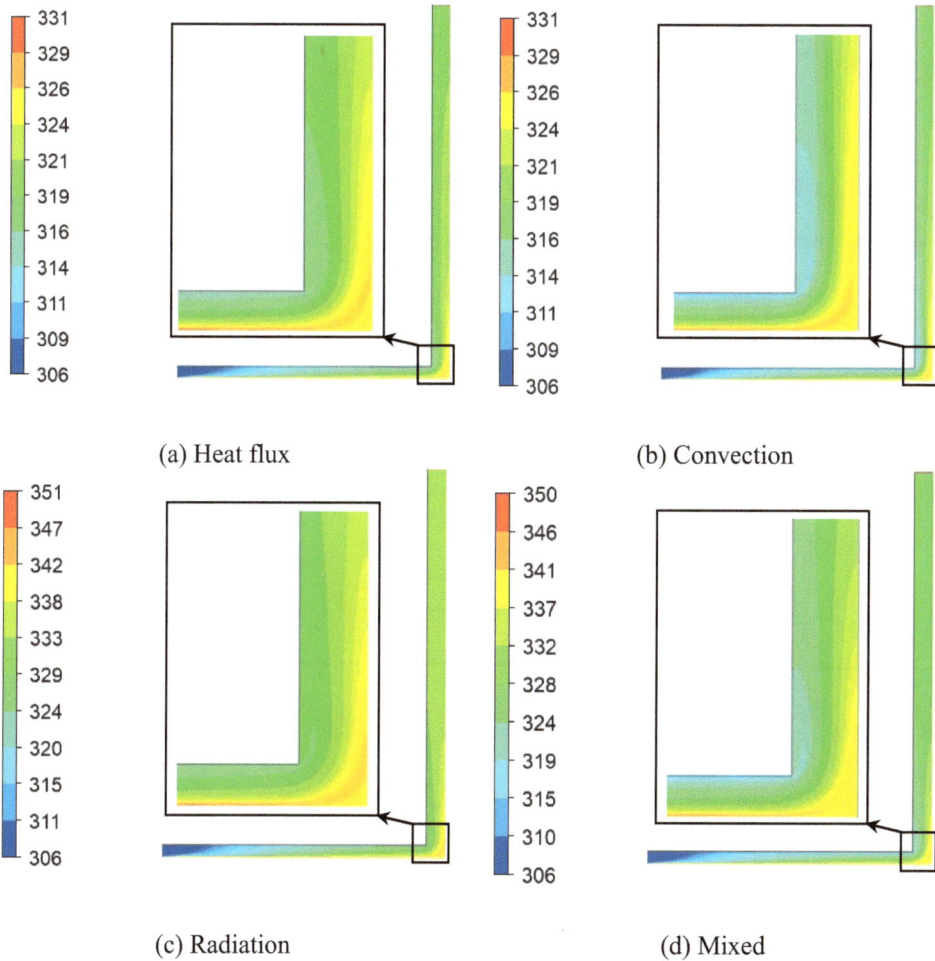

(a) Heat flux   (b) Convection

(c) Radiation   (d) Mixed

Fig. (3.27). Distribution of the temperature.

7.2. Magnitude Velocity Profiles

Fig. (**3.28**) shows the profiles of the magnitude velocity along with the collector for the four cases of heat transfer mode between the collector and the air. According to these results, it has been observed clearly that all cases are similar in the shape where the magnitude velocity increases in the collector until reaching the maximum value in the chimney inlet and remains constant along with the chimney. Besides, an acceleration zone has appeared in the chimney inlet. This zone was clearer in the first case corresponding to the convection.

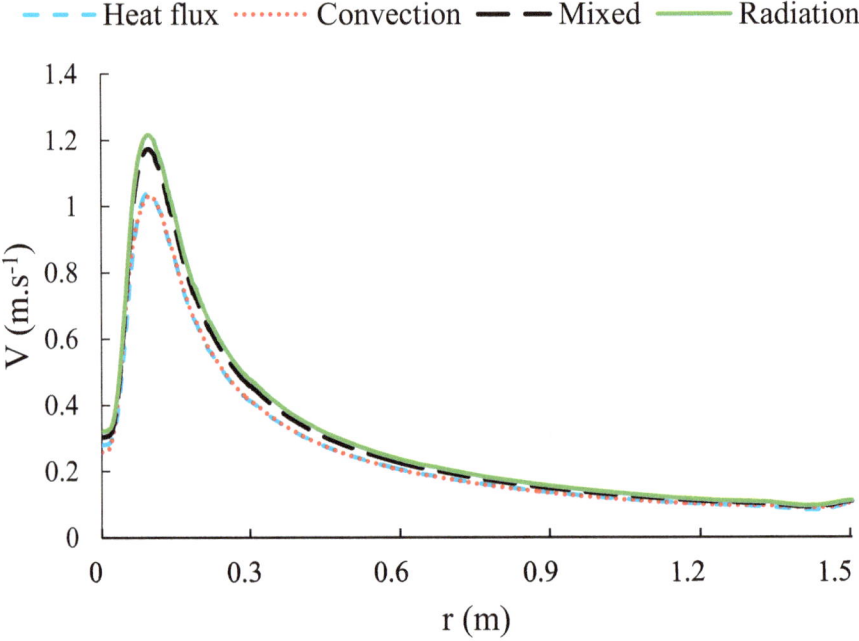

Fig. (3.28). Profiles of the magnitude velocity in the collector.

Fig. (**3.29**) presents the distribution of the magnitude velocity in the axisymmetric plane of the solar chimney for different heat transfer modes in the collector. According to these results, it has been noted that the velocity is weak in the inlet of the collector. Then, the magnitude velocity increases progressively in the collector until reaching the maximum value in the chimney inlet. In these conditions, the maximum value of the magnitude velocity is equal to $V=2.08$ m.s^{-1} with the heat flux mode. However, it is equal to $V=2.43$ m.s^{-1} with the radiation mode. Indeed, it has been observed that the acceleration zone is located in the base of the chimney. The increase of the temperature at the chimney base is due to the abrupt velocity change, under the conservation of the energy principle.

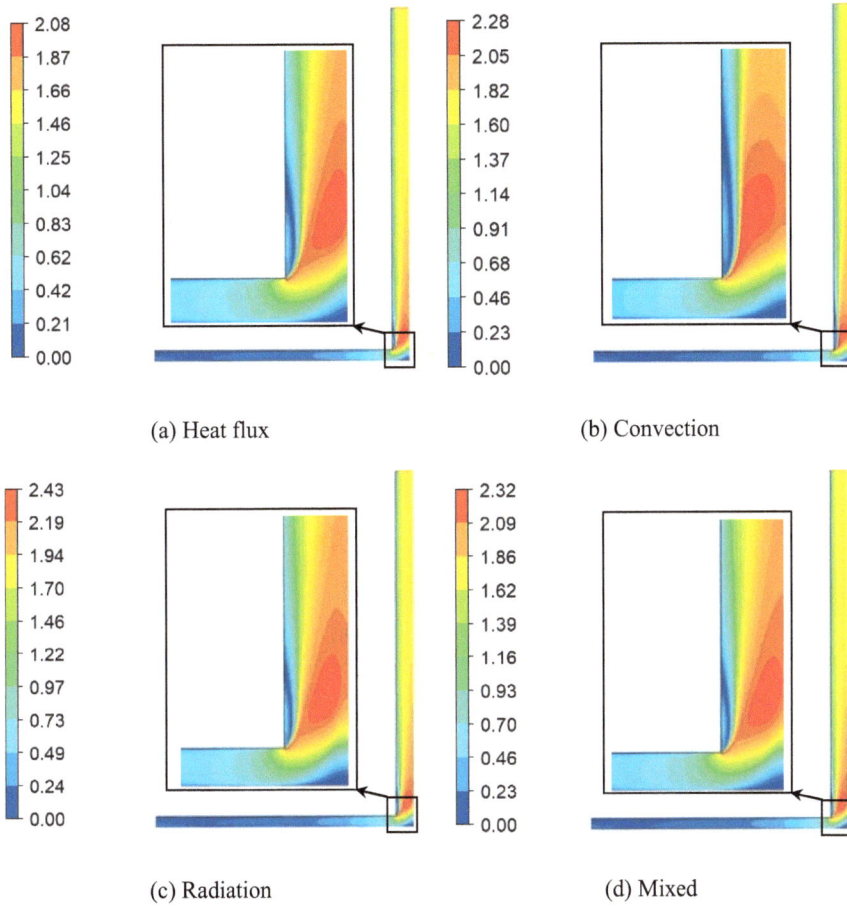

(a) Heat flux   (b) Convection

(c) Radiation   (d) Mixed

Fig. (3.29). Distribution of the magnitude velocity.

7.3. Radiation

The radiation is the principle parameter for the natural convection. From Fig. (**3.30**), it is clear that the radiation has the same distribution for all heat transfer modes. It is maximum on the collector because of the greenhouse effect and very weak in the chimney as this latter is isolated. The maximum radiation value is equal to $G=3270$ W.m^{-2} with the heat flux mode, to $G=3260$ W.m^{-2} with the convection mode, to $G=4460$ W.m^{-2} with the radiation mode and to $G=4450$ W.m^{-2} with the mixed mode.

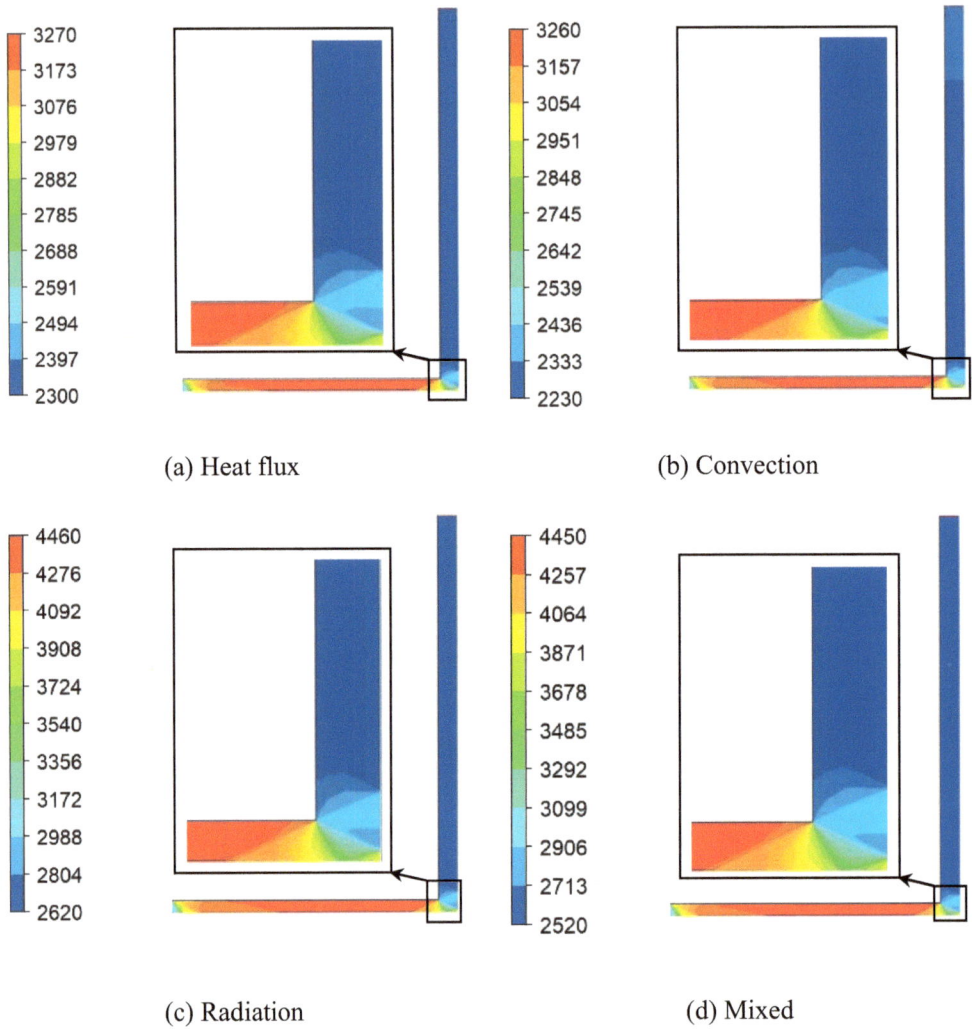

(a) Heat flux   (b) Convection

(c) Radiation   (d) Mixed

Fig. (3.30). Distribution of the radiation.

7.4. Enthalpy

Fig. (**3.31**) shows the distribution of the enthalpy in the axisymmetric plane of the solar chimney. According to these results, it is clear that the enthalpy distribution is similar to the temperature distribution for all heat transfer modes. The highest values are recorded on the absorber and in the axis of the chimney. Then, it decreases near the chimney walls due to the heat loss in the outside environment. The maximum enthalpy values are equal to h=32600 J.kg^{-1} with the heat flux mode, to h=32400 J.kg^{-1} with the convection mode, to h=52100 J.kg^{-1} with the radiation mode and to h=51500 J.kg^{-1} with the mixed mode.

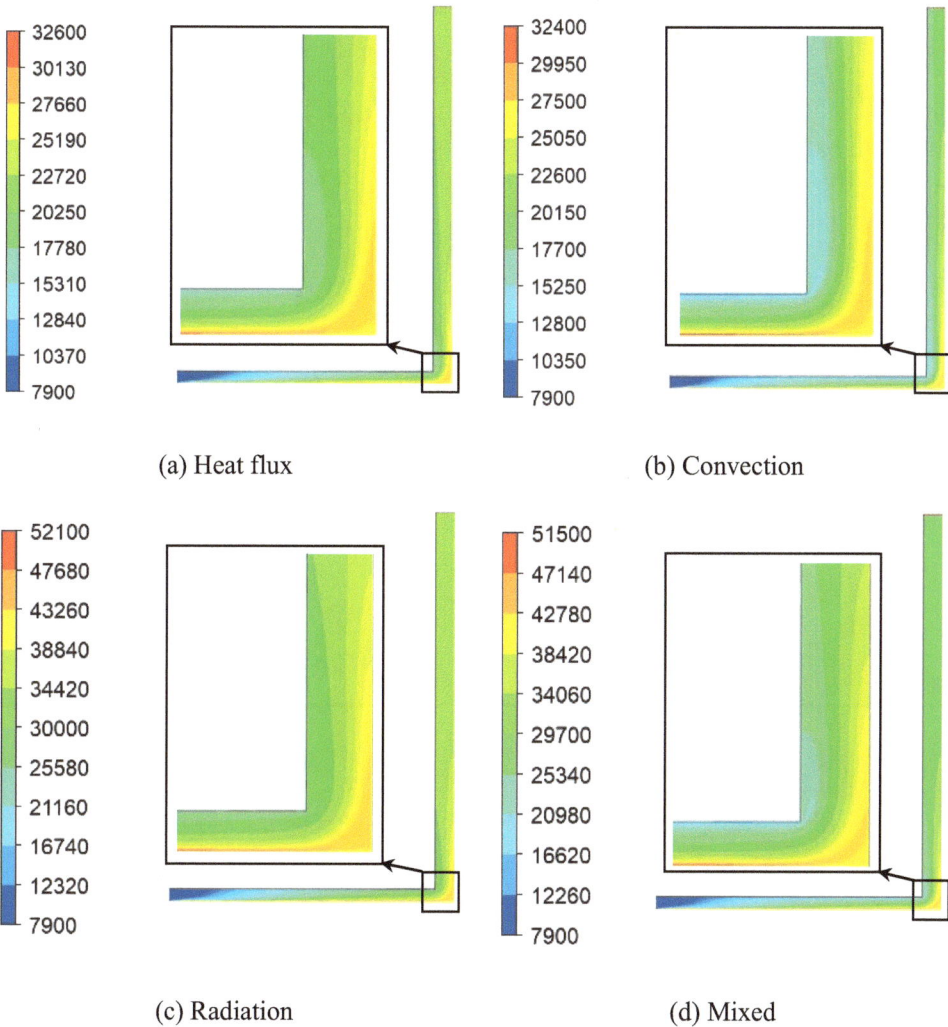

(a) Heat flux   (b) Convection

(c) Radiation   (d) Mixed

Fig. (3.31). Distribution of the enthalpy.

7.5. Static Pressure

Fig. (**3.31**) presents the distribution of the static pressure in the axisymmetric plane of the solar chimney for different heat transfer mode in the collector inlet. According to these results, it can easily be noted that the static pressure is on its maximum in the collector inlet. Besides, it has been observed a depression zone created in the base of the chimney. The static pressure continues to increase the way out of the chimney top. A brutal drop in the pressure has been noted just where the change of the airflow section. The comparison between the distribution of the magnitude velocity illustrated in Fig. (**3.29**) and the static pressure Fig. (**3.32**) confirms that the static pressure decreases with the increase of the velocity in all cases. The difference between the four models appears in the minimum value of the static pressure. The static pressure is equal to p= -3.28 Pa in the first case with the heat flux mode. However, it is equal to p= -4.44 Pa with the third case with the radiation mode.

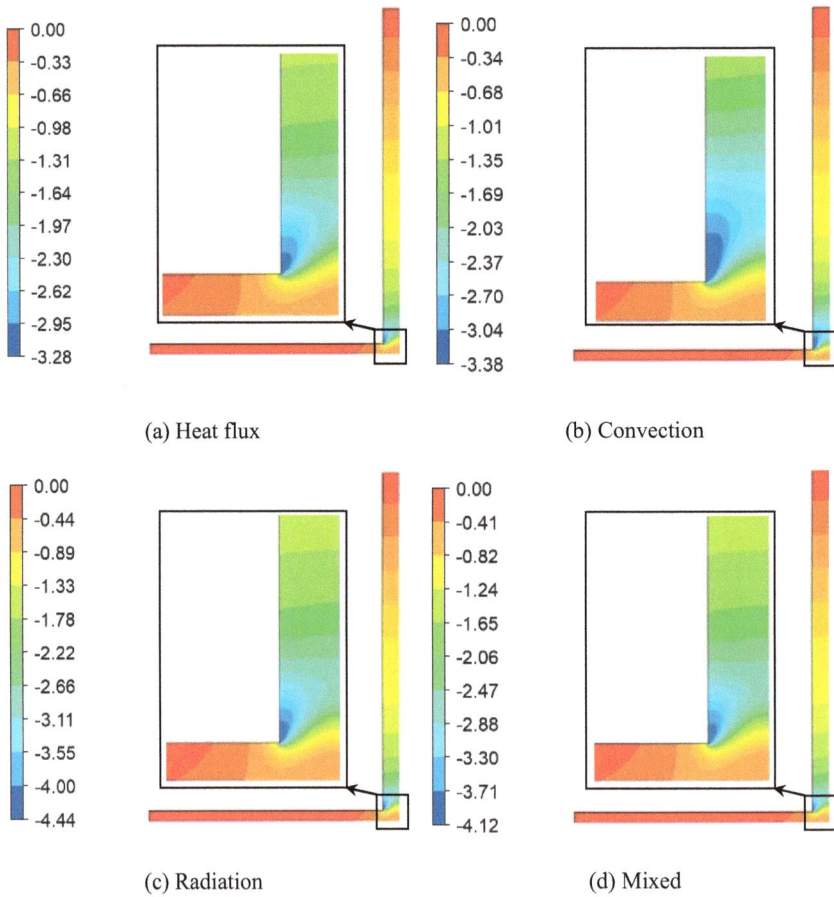

(a) Heat flux

(b) Convection

(c) Radiation

(d) Mixed

Fig. (3.32). Distribution of the static pressure.

7.6. Dynamic Pressure

Fig. (**3.33**) shows the distribution of the dynamic pressure in the axisymmetric plane of the solar chimney for different cases of the heat transfer modes. The dynamic pressure is the component of the fluid pressure that represents the fluid kinetic energy and it represents the dynamic effects of the flow. From these results, it has been noted that the dynamic pressure has approximately the same distribution of the magnitude velocity. The fact is that these two parameters are mathematically linked. Indeed, it has been observed that the dynamic pressure

reaches its maximum value near the chimney base and its value change for the different modes. The dynamic pressure value is equal to pd=2.44 Pa with the heat flux mode, to pd=2.92 Pa with the convection mode, to pd=3.30 Pa with the radiation mode, and pd=3.03 Pa with the mixed mode.

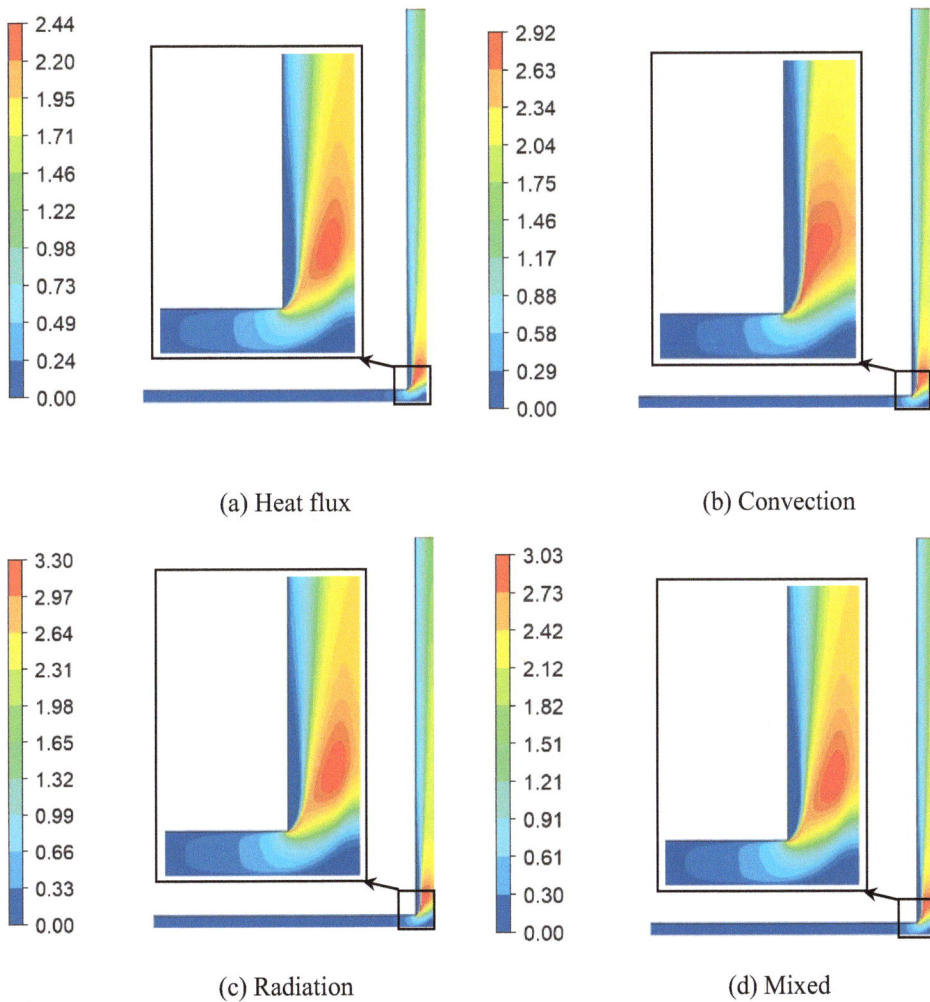

(a) Heat flux

(b) Convection

(c) Radiation

(d) Mixed

Fig. (3.33). Distribution of the dynamic pressure.

7.7. Turbulent Kinetic Energy

Fig. (**3.34**) presents the distribution of the turbulent kinetic energy in the axisymmetric plane of the solar chimney for different heat transfer mode. From these results, it has been noted that the turbulent kinetic energy is found to be very weak in the collector. The wake characteristic of the maximum value of the turbulent kinetic energy appears in the base of the chimney. The maximum value of the turbulent kinetic energy is found near the base of the chimney, where the change of the section of airflow. For example, the turbulent kinetic energy is equal to k= 0.281 m^2.s^{-2} with the convection mode, while it' is equal to k=3.58 m^2.s^{-2} with the mixed model.

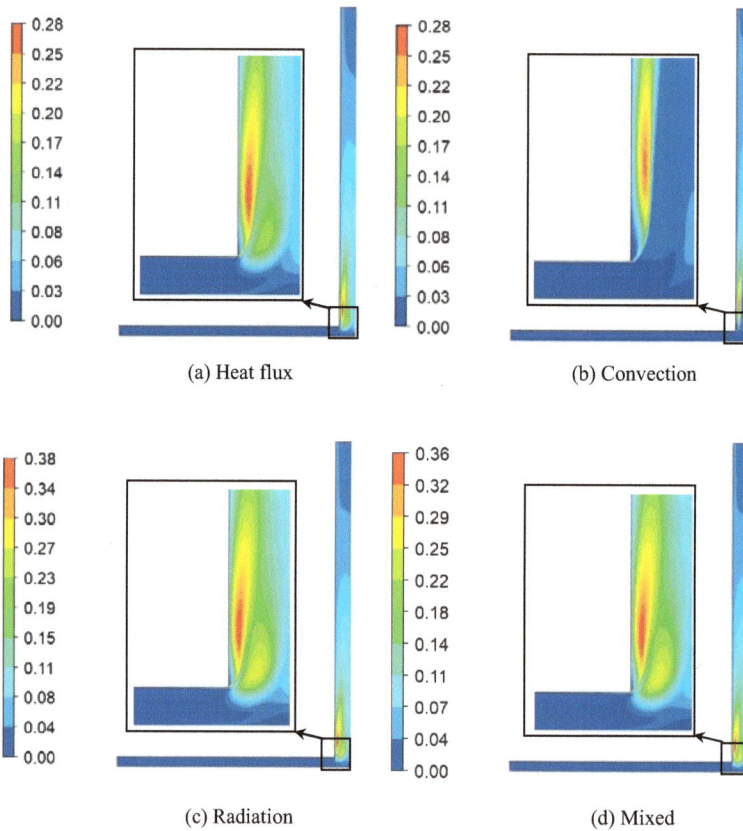

(a) Heat flux   (b) Convection

(c) Radiation   (d) Mixed

Fig. (3.34). Distribution of the turbulent kinetic energy.

7.8. Dissipation Rate of the Turbulent Kinetic Energy

Fig. (**3.35**) shows the distribution of the dissipation rate of the turbulent kinetic energy in the axisymmetric plane of the solar chimney system for the different heat transfer modes. According to these results, it has been observed that the dissipation rate of the turbulent kinetic energy in all cases is low, except for the collector output, which reaches its maximum owing to air friction on the collector. Indeed, the change of the heat transfer mode does not affect the distribution of the dissipation rate of the turbulent kinetic energy. The dissipation rate of the turbulent kinetic energy presents the same distribution for all heat transfer modes. The difference is illustrated at the maximum value which is equal to $\varepsilon=563$ m^2.s^{-3} with the heat flux mode, to $\varepsilon=565$ m^2.s^{-3} with the convection mode, to $\varepsilon=727$ m^2.s^{-3} with the radiation mode and $\varepsilon=679$ m^2.s^{-3} with the mixed mode.

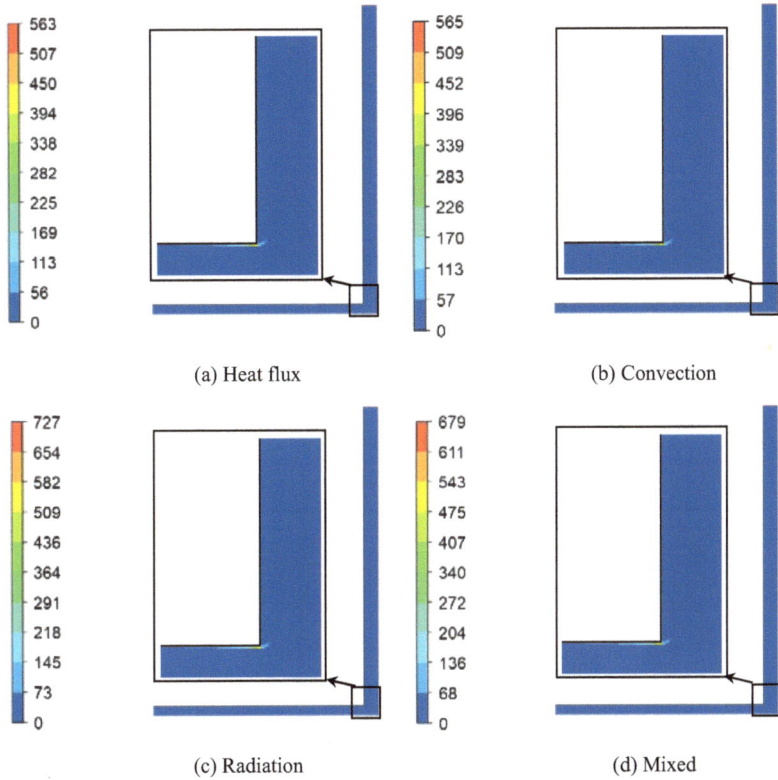

(a) Heat flux   (b) Convection

(c) Radiation   (d) Mixed

Fig. (3.35). Distribution of the dissipation rate of the turbulent kinetic energy.

CONCLUSION

Many encouraging studies were reported about the use of a solar chimney to produce electrical power. Performance studies were related to numerical parameters. In this chapter, we are managed to perform a numerical model approaching reality and compare it with the experimental results of Kasaeian *et al.* (2014). The comparison of our numerical results with experimental results shows a good agreement and confirms the numerical method. In the first step, we have concluded that the best model should have the biggest number of cells. In our case, the maximum number of cells was 326 105 cells. In the second step, we have studied the turbulence models effect and we have compared our numerical results with the experimental data. In these conditions, we have found that the realizable k-ε is the most adequate turbulent model. In the third step, we have studied the effect of heat transfer mode. Particularly, we are interested in the heat transfer in the absorber and the collector. As a result, the best model is found with the convection mode in the two cases.

CHAPTER 4

Design of Prototype

1. INTRODUCTION

Numerous studies about the use of solar chimney for air conditioning, drying, and electrical power production were performed. The present study considered first the heat transfer process and the fluid dynamics in the collector and the chimney tower and extended later by a parametric study on the effects of the geometrical parameters. A validated CFD code was adopted to the solar chimney shape to resolve the governing equations (continuity, momentum, and energy equations). The investigation displays a necessity for a more detailed analysis of such systems, which is essential for an ample definition of the design rules. The available literature is scarce on this type of analysis, as research mostly concentrates on the evaluation of the global performance of such systems. For the design of real systems, more detailed studies of the geometrical and operational aspects are needed, involving meteorological conditions and geometric performances. This work is based on understanding the mechanisms of flow and transfer of natural convection in two-dimensional configurations using the CFD software ANSYS Fluent to find the optimal design that allows adequate thermal control and maximum energy performance. Besides, we have studied numerically the effect of the geometrical parameters, such as the chimney diameter, the chimney height, the collector diameter, the collector slope angle, and the collector height, thus to obtain the optimal parameters.

2. SOLAR CHIMNEY SYSTEM

The solar chimney prototype is composed of three main parts, the chimney, the collector, and the absorber (Fig. **4.1**). The chimney is a PVC pipe, its height and diameter are H and d respectively with a thickness of 3 mm, the absorber is a wooden platform of diameter D with an 8 mm thickness painter with a black layer. The wooden platform was chosen because it is a good thermal insulator and causes to improve the performance of the system. We choose polyethylene low density such a collector roof with a thickness of 1 mm, h indicates the inlet height, and θ indicates the collector slope angle. Different proprieties of our materiel are given in Table **4.1**. The geometrical arrangements are illustrated in Fig. (**4.2**).

Haythem Nasraoui, Moubarek Bsisa & Zied Driss

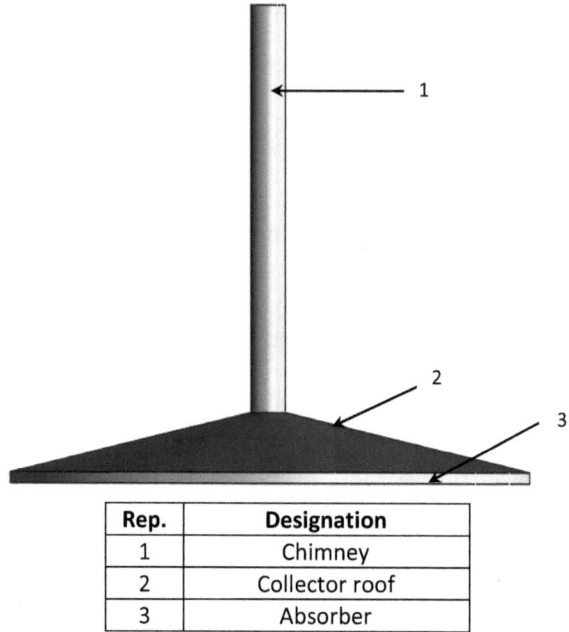

Rep.	Designation
1	Chimney
2	Collector roof
3	Absorber

Fig. (4.1). 3D model of the prototype.

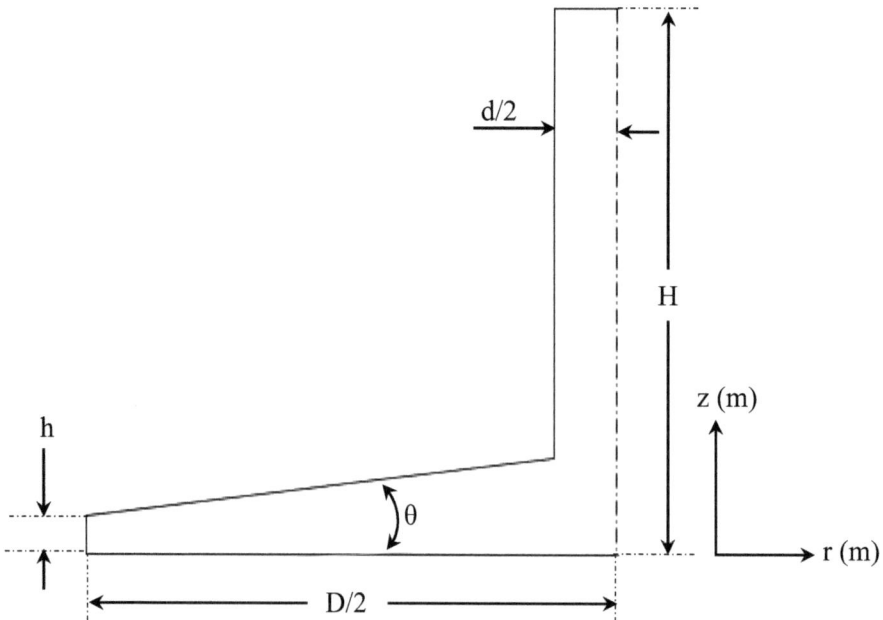

Fig. (4.2). Two-dimensional axisymmetric model of prototype.

Table 4.1. Characteristics of materials.

	Material	Density ρ (kg.m^{-3})	Specific Heat Cp (J.kg^{-1}.k^{-1})	Thermal Conductivity K (W.m^{-1}.k^{-1})	Absorption Coefficient α (m^{-1})
Fluid	Air	1.122	1006.43	0.0242	0
Solid	Wood	700	2310	0.024	0.8
	PVC	1380	1046	0.2	0.6
	Polyethylene	920	2100	0.33	0.4

3. NUMERICAL PARAMETERS

3.1. Meshing

Meshing is one of the most important aspects of any CFD simulation. According to the study of the meshing effect in the last chapter, we use in this chapter, the mesh presenting 356 000 cells. The schematic of the meshing is shown in Fig. (**4.3**).

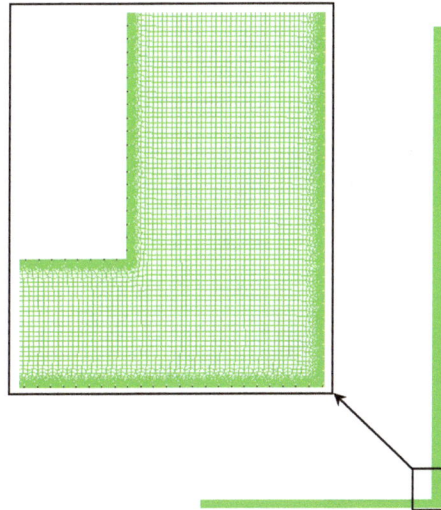

Fig. (4.3). Meshing.

3.2. Boundary Conditions and Numerical Parameters

In this chapter, a general description was given based on the different models that we fixed in chapter 2. A physical model was simulated using ANSYS Fluent 17.0, based on the geometrical dimensions of the axisymmetric solar chimney. The boundary conditions used are shown in Fig. (**4.4**). The wall boundary is used for the chimney, the collector, and the absorber. The insulated wall was applied for the surface of the chimney and convective heat transfer option was applied for different parts of the device such as collector and absorber. For the outlet and the inlet of the system, static pressure takes a value of zero, which means that at this opening, the air enters, and exits the system to an area of static atmospheric pressure. The governing equations were solved, assuming symmetric and steady-state conditions. The realizable k-ε turbulent model is applied to describe turbulent flow conditions and DO model is used as a solar radiation model. A second-order upwind discretization is used for the momentum equation and turbulent kinetic energy and specific dissipation rate. Table **4.2** shows the different boundary conditions parameters.

Table 4.2. Boundary condition parameters.

Absorber	Wall and convection	$h = 8$ W.m^{-2}.K^{-1}
Collector roof	Wall and convection	$h = 8$ W.m^{-2}.K^{-1}
Chimney surface	Wall and heat flux	$q = 0$ W.m^{-2}
Collector Inlet	Pressure inlet	$P_i = 0$ Pa, $T_a = 306$ K
Chimney outlet	Pressure outlet	$P_o = 0$ Pa

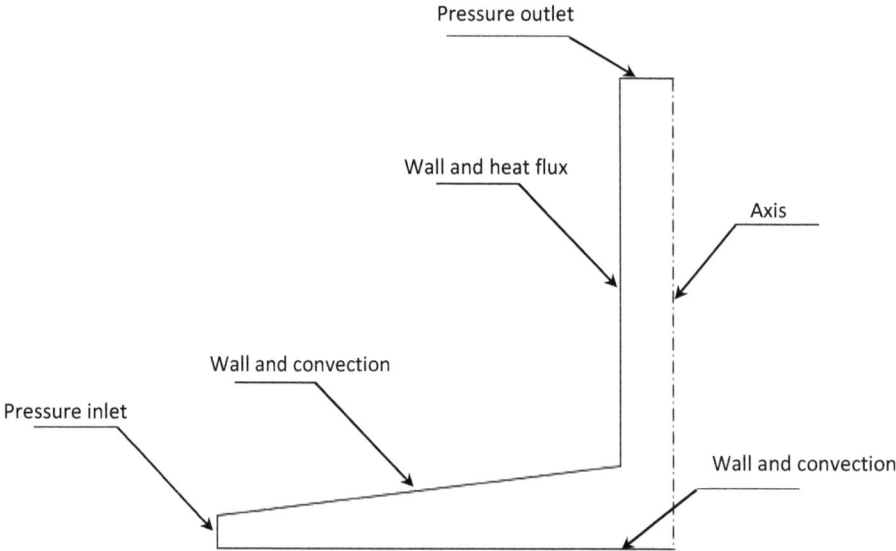

Fig. (4.4). Boundary conditions.

4. SIZING OF PROTOTYPE

In this section, we have presented the steps that we have proceeded to choose the optimal size of our prototype. We have varied many dimensions, such as the collector height, the chimney diameter, the chimney height, the collector slope angle (angle θ), and the collector diameter of the prototype. 88 simulations have been launched to obtain optimal prototype dimensions. We take four chimney heights, for each height is simulated four chimney diameters and each diameter is simulated four collector height. The value of simulations parameters is illustrated in Table **4.3**. Besides, appendix 1 gives the maximum velocity values.

All simulations will simulate in the same condition to be comparing between their results (for April 20, 2016 at 12:00 pm). Based on the forecast of weather data in Sfax, the solar radiation for April 20, 2016 at 12:00 pm is assumed to equal $G=927$ W.m^2. The solar radiation consists of 746 W.m^2 of direct radiation and 181 W.m^2 of diffuse radiation with an ambient temperature of T=306 K. The choice of optimal geometry is based on the calculation of the maximum velocity value, the distribution magnitude velocity, the static pressure and the temperature inside the prototype.

Table 4.3. Simulation parameters.

Parameters	Values			
Collector diameter D	2 m	2.75 m	3.5 m	4 m
Collector slope angle θ	0°	5°	10°	15°
Collector height h	0.005 m	0.05 m	0.075 m	0.1 m
Chimney diameter d	0.1 m	0.13 m	0.16 m	0.2 m
Chimney height H	1 m	2 m	3 m	4 m

4.1. Collector Diameter Effect

In this section, we are interested in the study of the collector diameter effect on the behavior of the flow. The local characteristics such as magnitude velocity, temperature, and static pressure are evaluated for different values of the diameter equal to D=2 m, D=2.75 m, D=3.5 m, and D=4 m. In these conditions, the other parameters are defined by H=3 m, h=0.05 m, d=0.16 m, and θ=0°. The schematic of the system is shown in Fig. (**4.5**) and the different geometries are shown in Fig. (**4.6**).

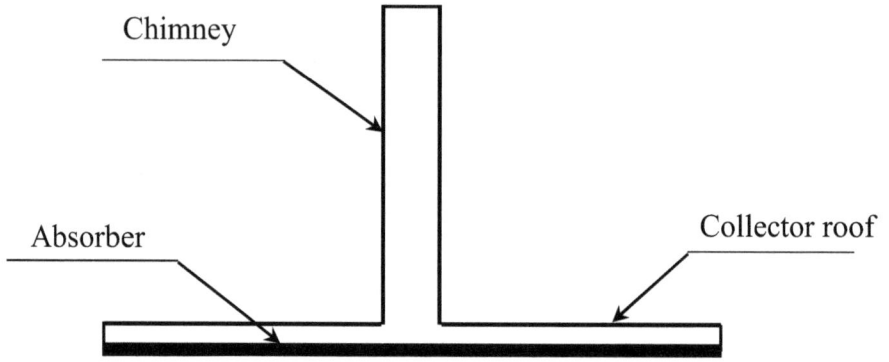

Fig. (4.5). Schematic of the system.

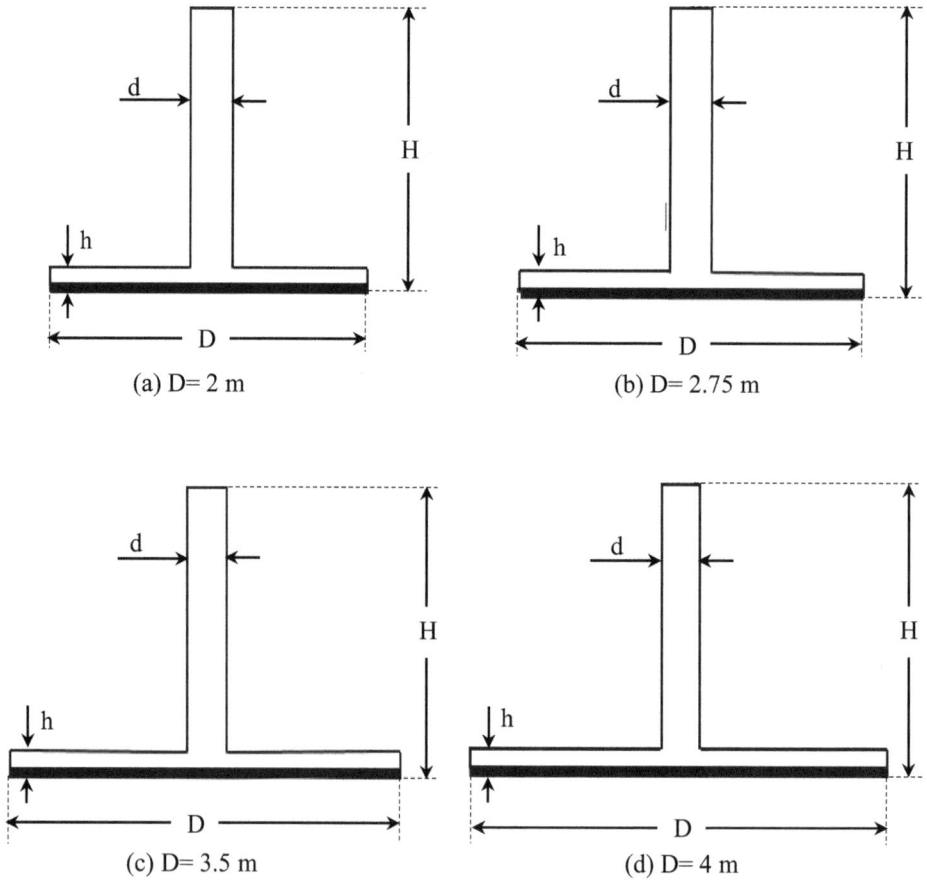

(a) D= 2 m

(b) D= 2.75 m

(c) D= 3.5 m

(d) D= 4 m

Fig. (4.6). Schematic of the different geometries.

4.1.1. Magnitude Velocity

Fig. (**4.7**) shows the variation of the maximum magnitude velocity value for different collector diameter. According to these results, it is clear that the maximum value of the magnitude velocity increases with the increase in the collector diameter. The elevation of the collector area means that the difference in the temperature between inlet and outlet collector increases. The theoretical maximum magnitude velocity is esteemed by Unger (1988) in equation (1-16) and it is proportional to the temperature difference. Therefore, we noticed that an increase in the collector diameter increases also the maximum value of the magnitude velocity. On the practical level, it is preferable to increase the collector area.

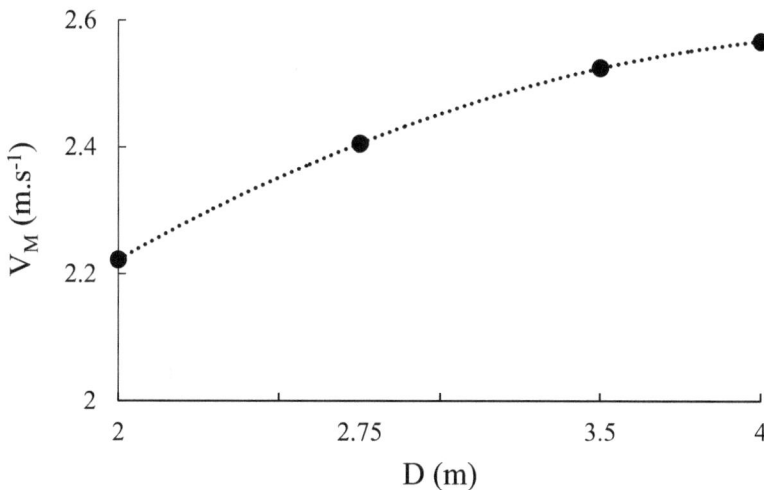

Fig. (4.7). Variation of the maximum magnitude velocity.

Figs. (**4.8**) and (**4.9**) illustrate the profiles of the magnitude velocity along the chimney axis and in the collector. According to these results, it is clear that the variation of the collector diameter does not affect the shape of the magnitude velocity profile. The magnitude velocity is minimal at the absorber where $z=0$ m and reaches his maximum value equal to $V= 2.4$ m.s^{-1}, at the chimney inlet where $z=0.05$ m. After that, the magnitude velocity decreases until the position $z=1$ m where the velocity maintains a constant value until the chimney outlet. The influence of the collector diameter on the magnitude velocity has also been observed along with the chimney. In these conditions, the magnitude velocity increases with the diameter. In the collector, the velocity has the minimum value at the collector inlet and it increases until the collector outlet where it reaches its maximum value equal to V=1.22 m.s^{-1} due to the elevation of the greenhouse effect in the collector. The difference between the four profiles is not important except for the maximum magnitude velocity value in the collector outlet.

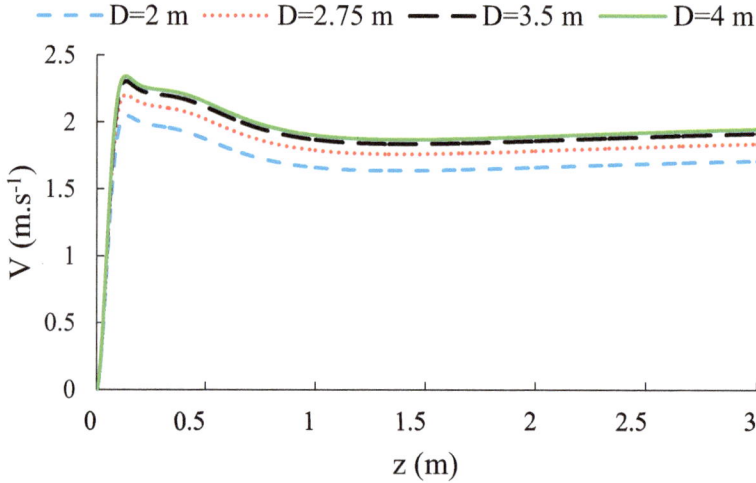

Fig. (4.8). Profiles of the magnitude velocity along the chimney axis.

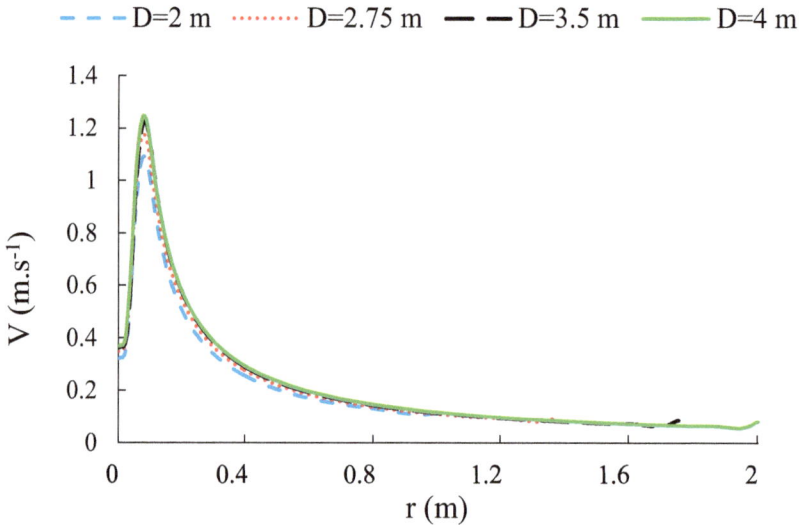

Fig. (4.9). Profiles of the magnitude velocity in the collector.

Figs. (**4.10**) and (**4.11**) show the distribution of the velocity fields and the magnitude velocity in the axisymmetric plane of the solar chimney for the different collector diameters equal to D=2 m, D=2.75 m, D=3 m, and D=4 m. According to these results, the velocity conserves the same distribution for all diameters. The acceleration zone is located near the chimney inlet. In the collector inlet, the velocity is very low because it does not yet heat. These results confirm also that the velocity fields present a uniform distribution along with the chimney except at the inlet. Indeed, it is clear that the chimney diameter has a direct effect on the maximum value of the magnitude velocity value equal to V=2.22 m.s^{-1} with a diameter D=2 m, to V=2.41 m.s^{-1} with diameter D=2.75 m, to V=2.52 m.s^{-1} with diameter D=3.5 m and to V=2.57 m.s^{-1} with diameter D=4 m. These results confirm that the maximum value of the magnitude velocity increases with the increase in the chimney diameter.

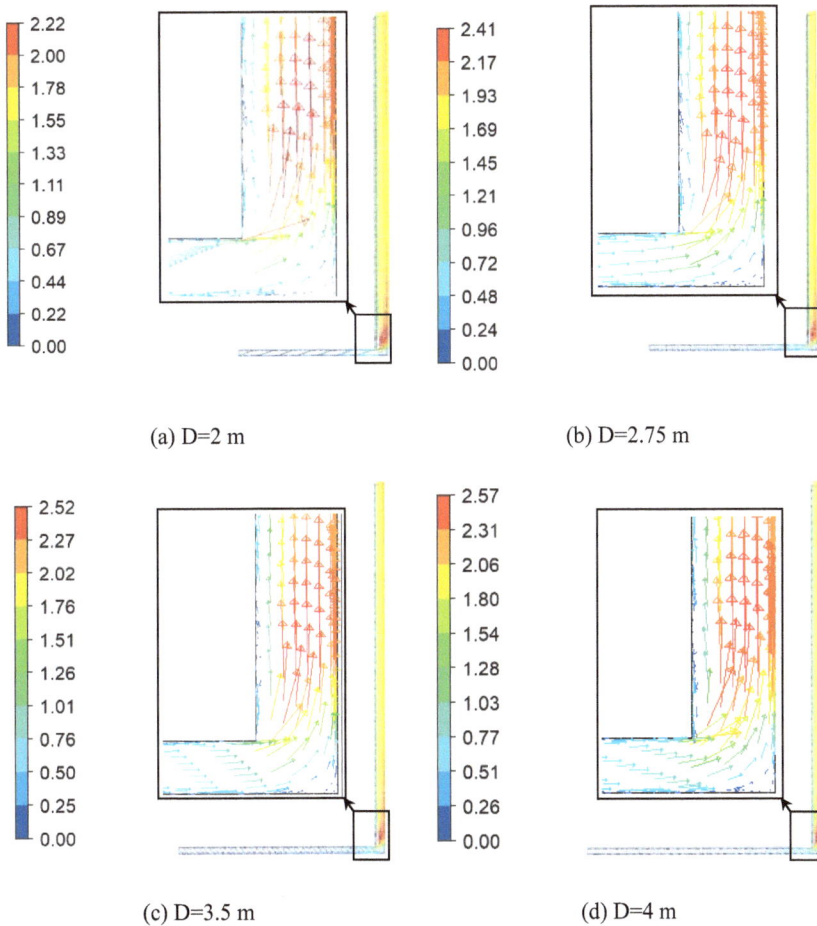

(a) D=2 m   (b) D=2.75 m

(c) D=3.5 m   (d) D=4 m

Fig. (4.10). Velocity fields for different collector diameters.

(a) D=2 m   (b) D=2.75 m

(c) D=3.5 m   (d) D=4 m

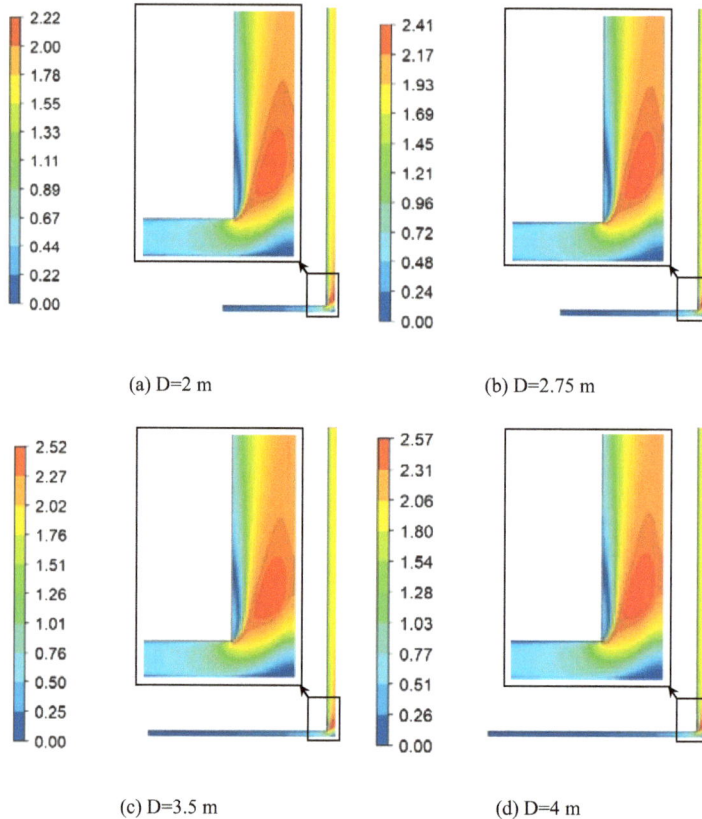

Fig. (4.11). Distribution of the magnitude velocity for different collector diameters.

4.1.2. Temperature

Fig. **(4.12)** displays the temperature profiles in the collector region. According to these results, it is clear that the increase of the collector diameter gives more convection with air. For this, a temperature elevation in the absorber has been observed, which directly affects the performance of the chimney. Fig. **(4.13)** illustrates the distribution of the temperature in the axisymmetric plane of the solar chimney for different collector diameters. According to these results, it has been observed that the maximum value of the temperature is localized in the absorber because its absorption coefficient is height. Through the greenhouse, the air becomes heated when the diameter is the largest. This fact can be explained by the floatability phenomenon of air. Indeed, it is clear that the difference between the maximum temperature values is not very important. However, the difference is localized in the temperature distribution around the chimney axis. In fact, for a low diameter of the collector, it is clear that the hot air flow is limited to the axis of the chimney. However, for the other diameters, the hot air distribution up to the chimney walls since the flow becomes more important.

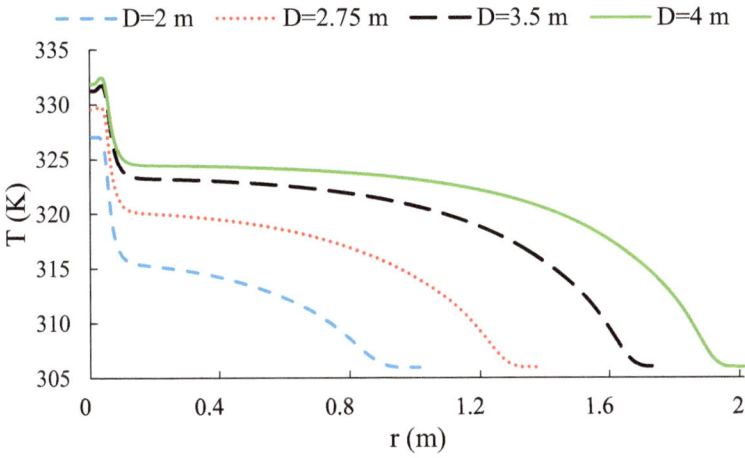

Fig. (4.12). Profiles of the temperature in the collector.

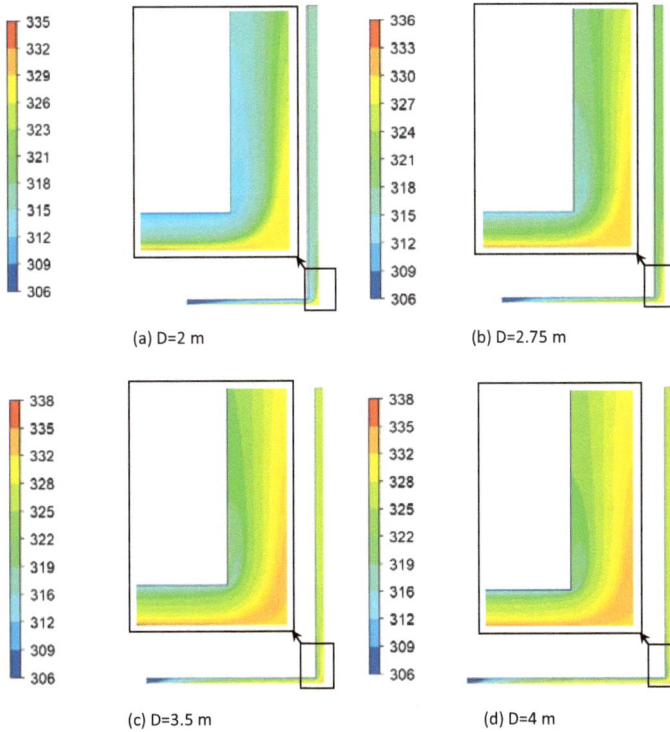

Fig. (4.13). Distribution of the temperature for different collector diameters.

4.1.3. Static Pressure

Fig. (**4.14**) proves the profiles of the static pressure along with the chimney for the different diameters of the collector equal to D=2 m, D=2.75 m, D=3 m, and D=4 m. From these results, it is clear that the static pressure presents similar distribution for all cases. The depression value is high at the chimney entrance, where the velocity is maximum. Then, the static pressure rises and reaches the atmospheric pressure at the chimney exit. Otherwise, this pressure difference is liable to the airflow along with the chimney. Besides, it has been observed that the increase in the collector diameter leads to the increase in the depression at the base of the chimney. This fact is illustrated from the static pressure distribution presented in the axisymmetric plane of the chimney axis for different diameters of the collector as shown in Fig. (**4.15**). According to these results, it is clear that the depression zone is located in the base of the chimney. Elsewhere, the static pressure increases until reaching the atmospheric pressure in the inlet of the collector and the outlet of the chimney. In these conditions, the difference is clear at the maximum value equal to p=3.84 Pa with a diameter D=2 m, to p=4.47 Pa with D=2.75 m, to p=4.86 Pa with D=3.5 m and to p=5.05 Pa with D=4 m.

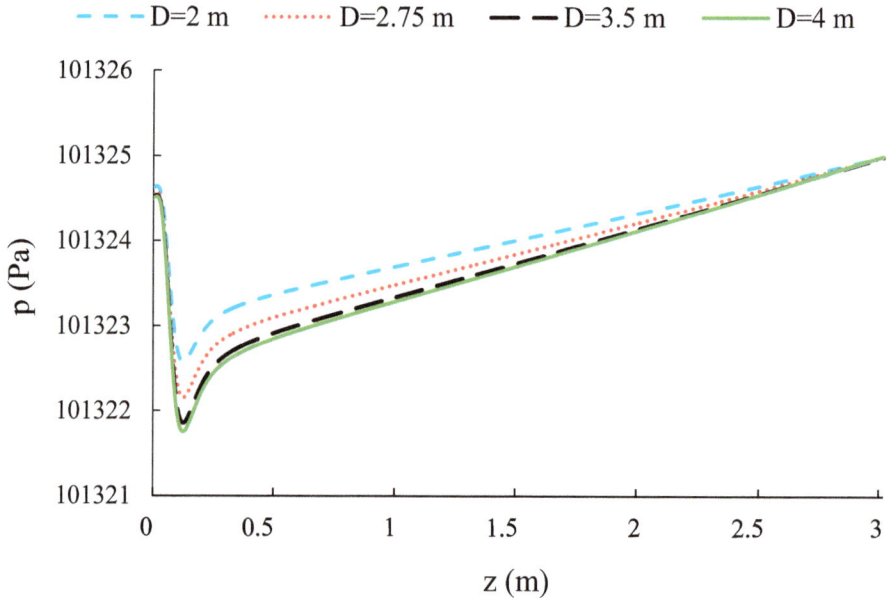

Fig. (4.14). Profiles of the static pressure along the chimney axis.

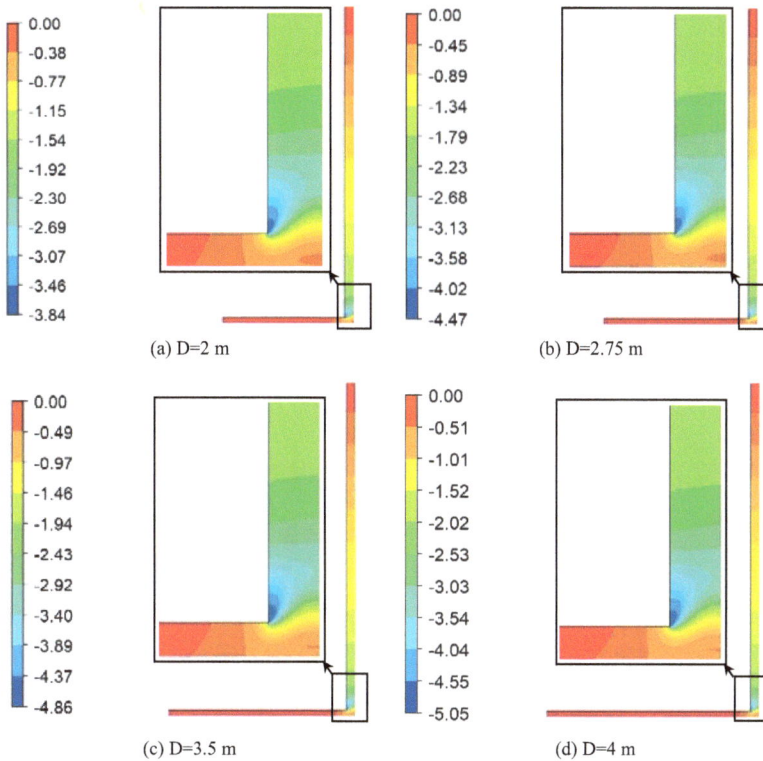

(a) D=2 m   (b) D=2.75 m

(c) D=3.5 m   (d) D=4 m

Fig. (4.15). Distribution of the static pressure for different collector diameters.

4.1.4. Turbulent Kinetic Energy

Fig. (**4.16**) displays the distribution of the turbulent kinetic energy in the axisymmetric plane of the solar chimney for the different collector diameters. According to these results, the turbulent kinetic energy presents identical distribution for all cases. The turbulent kinetic energy is found to be very weak in the axisymmetric plane except in the chimney inlet. Also, there may be a small difference between the distributions of the turbulent kinetic energy for each diameter. This difference appears in the chimney inlet when the highest values are found in the fourth case with diameter D=4 m. The maximum value of the turbulent kinetic energy is equal to $k=0.314$ $m^2.s^{-2}$ with a diameter D=2 m, to $k=0.367$ $m^2.s^{-2}$ with a diameter D=2.75 m, to $k=0.403$ $m^2.s^{-2}$ with a diameter D=3.5 m and to $k=0.415$ $m^2.s^{-2}$ with a diameter D=4 m.

(a) D=2 m   (b) D=2.75 m

(c) D=3.5 m   (d) D=4 m

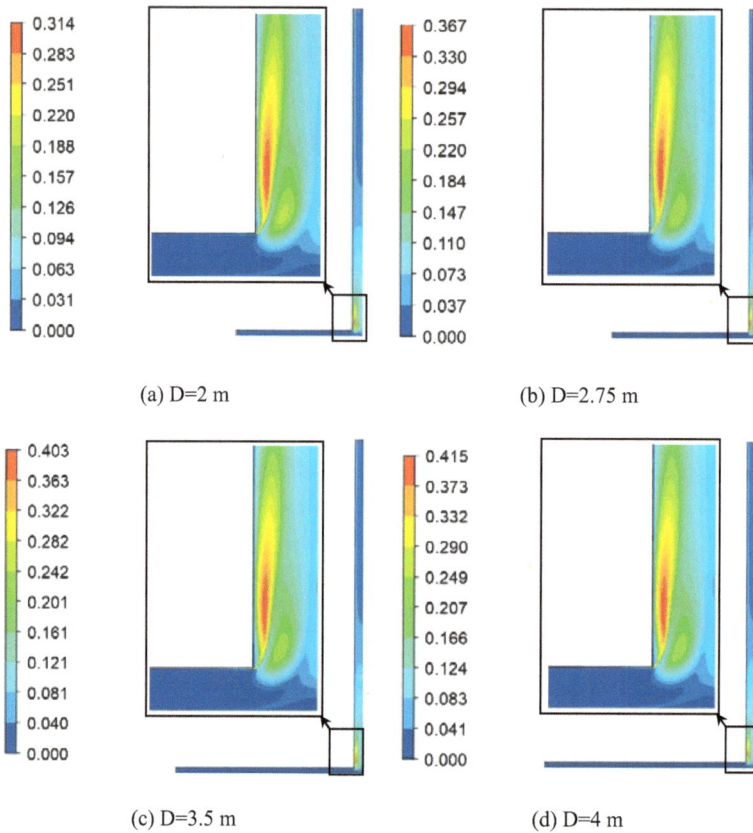

Fig. (4.16). Distribution of the turbulent kinetic energy for different collector diameters.

4.1.5. Dissipation Rate of the Turbulent Kinetic Energy

The dissipation phenomena are due to the turbulent kinetic energy when it was converted to internal energy for example, the thermal energy. Fig. (**4.17**) shows the distribution of the dissipation rate of the turbulent kinetic energy in the axisymmetric plane of the solar chimney for the different collector diameters equal to D=2 m, D=2.75 m, D=3.5 m, and D=4 m. According to these results, the variation of the collector diameter does not affect the distribution of the dissipation rate of the turbulent kinetic energy, which is found to be very weak in the axisymmetric plane except in the collector outlet. Indeed, it is clear that the dissipation rate of the turbulent kinetic energy increases with the increase of the collector diameter. The difference appears at the maximum value, which is equal to ε=337 $m^2.s^{-3}$ with a diameter D=2 m, to ε=368 $m^2.s^{-3}$ with a diameter D=2.75 m, to ε=407 $m^2.s^{-3}$ with a diameter D= 3.5 m and to ε=410 $m^2.s^{-3}$ with a diameter D=4 m.

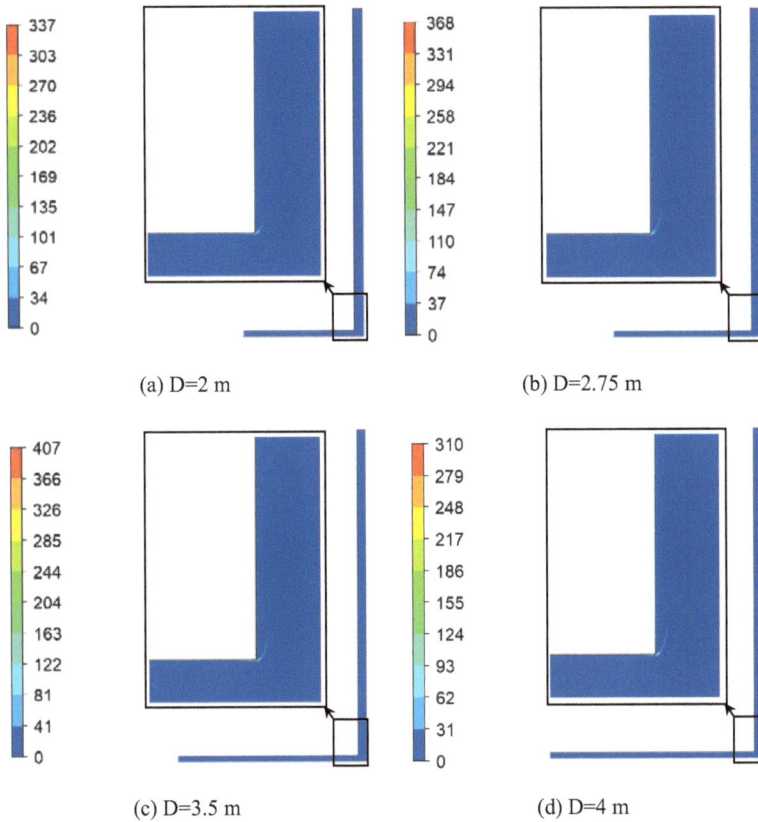

(a) D=2 m   (b) D=2.75 m

(c) D=3.5 m   (d) D=4 m

Fig. (4.17). Distribution of the dissipation rate of the turbulent kinetic energy for different collector diameters.

4.1.6. Turbulent Viscosity

Fig. (**4.18**) shows the distribution of the turbulent viscosity in the axisymmetric plane of the solar chimney for the different collector diameters equal to D=2 m, D=2.75 m, D=3.5 m, and D=4 m. According to these results, the turbulent viscosity presents the same distribution for all diameters. Indeed, it is clear that the turbulent viscosity is found to be very weak in the collector, but it is maximum at the chimney inlet. This fact can be explained by the collision between the particles constituting the volume of air. Besides, there may be a small difference between the distributions of the turbulent kinetic energy for each diameter. This difference is located in the chimney inlet when the highest value is found in the fourth case with a diameter D=4 m. Indeed, it has been noted that the turbulent viscosity increases with the increase of the collector diameter. In fact, the maximum value of the turbulent viscosity is equal to μ_t =0.00247 kg.m^{-1}.s^{-1}

with a diameter D=2 m, μ_t =0.0026 kg.m^{-1}.s^{-1} with a diameter D=2.75 m, μ_t =0.00268 kg.m^{-1}.s^{-1} with diameter D=3.5 m and μ_t =0.0027 kg.m^{-1}.s^{-1} with diameter D=4 m.

(a) D=2 m   (b) D=2.75 m

(c) D=3.5 m   (d) D=4 m

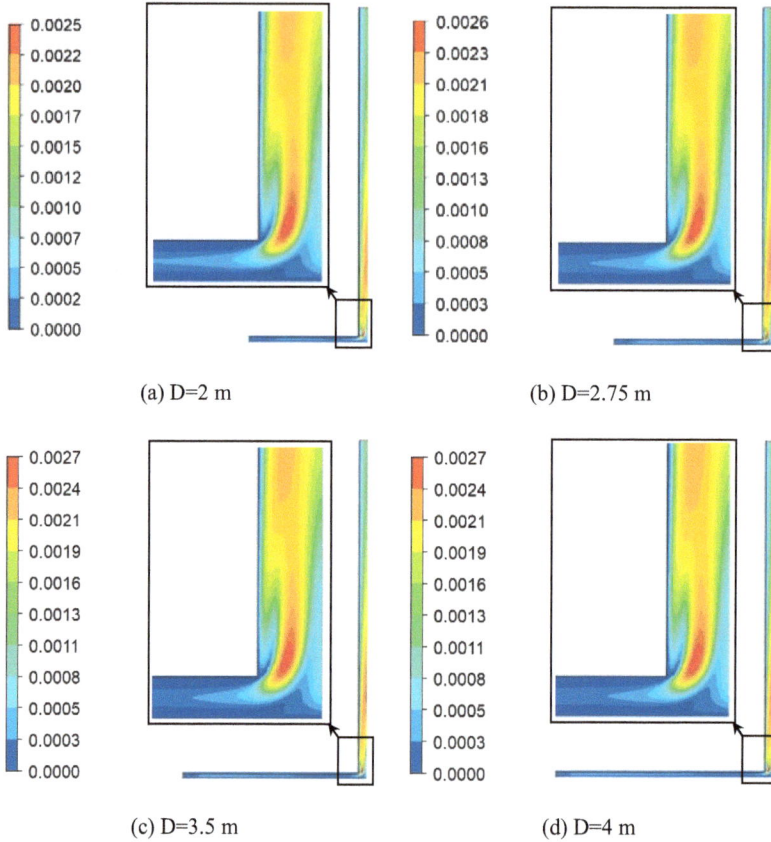

Fig. (4.18). Distribution of the turbulent viscosity for different collector diameters.

4.2. Collector Slope Angle Effect

In this section, we are interested in the study of the collector slope angle effect on the local results, such as the magnitude velocity, the temperature, the static pressure, and the turbulence characteristics. The considered system is defined by a collector height equal to h=0.005 m, a chimney diameter D=0.16 m, a collector diameter D=2.75 m and a chimney height H=3 m (Fig. **4.19**). Particularly, we have studied the solar chimney for four collector slope angles equal to θ=0°, θ=5°, θ=10°, and θ=15°, as shown in Fig. (**4.20**).

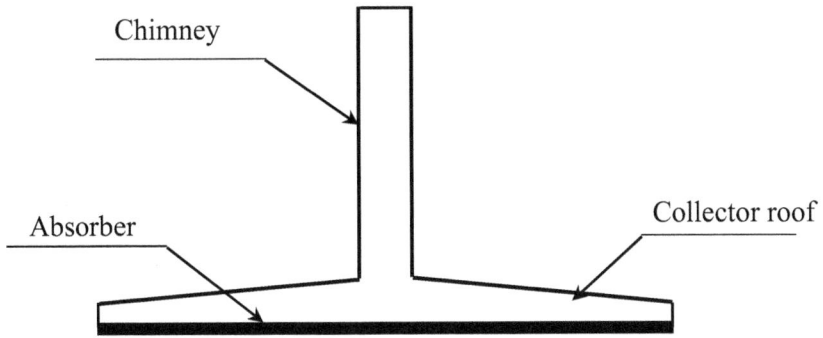

Fig. (4.19). Schematic of the system.

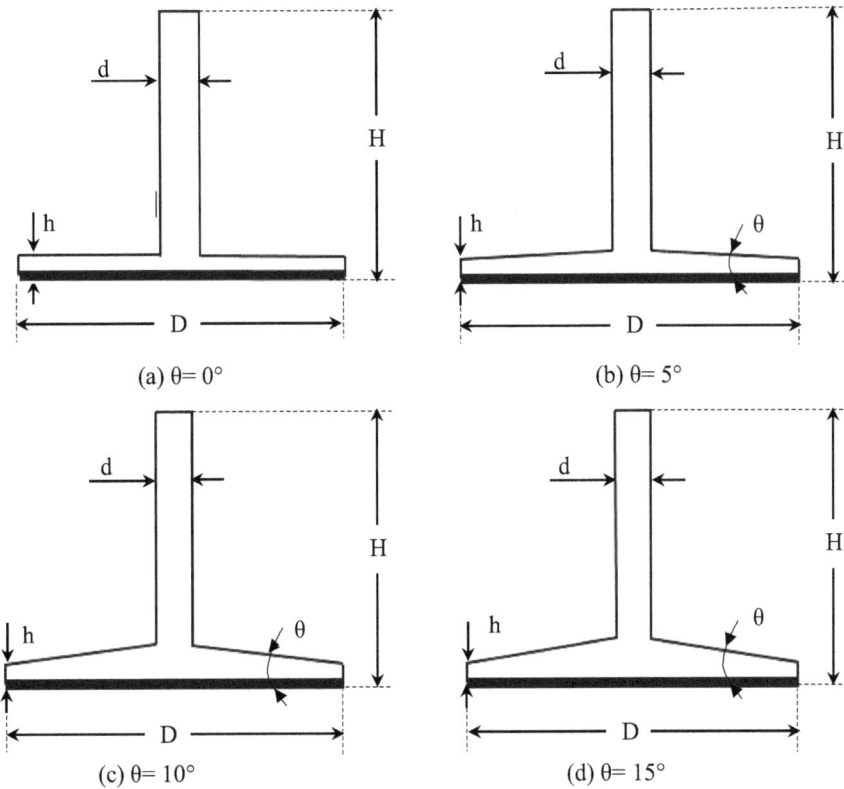

(a) θ= 0°

(b) θ= 5°

(c) θ= 10°

(d) θ= 15°

Fig. (4.20). Schematic of the different geometries.

4.2.1. Magnitude Velocity

Fig. (**4.21**) presents the profiles of the magnitude velocity along the chimney axis for different collector slope angles. According to these results, these profiles present the same variation. The maximum value appears at the chimney inlet. Then, it remains lightly uniform along with the chimney and particularly in the outlet. The comparison between these profiles confirms that the maximum value of the magnitude velocity appears with a collector angle θ=0°. In fact, for the other cases, the air volume increases and the difference in temperature in the collector decreases. Indeed, it has been noted that the maximum value of the magnitude velocity is directly proportional to the difference in the temperature. For this, we propose to choose the collector slope angle of θ=0°.

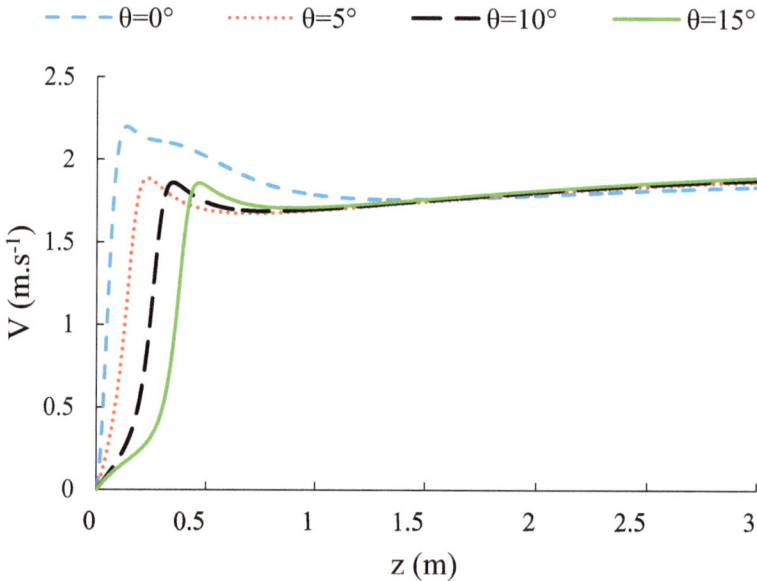

Fig. (4.21). Profiles of the magnitude velocity along the chimney axis.

The distribution of the velocity fields and the magnitude velocity in the axisymmetric plane of the solar chimney for different collector slope angle angles equal to θ=0°, θ=5°, θ=10°, and θ=15° is shown, respectively in Figs. (**4.22**) and (**4.23**). According to these results, the accelerated zones are located at the entrance of the chimney for the angle zero and the chimney outlet for the other slope angles. Moreover, the maximum velocity value is in the first case because of the airflow in the collector with this angle move in the straight streamlines shown in Fig. (**4.22**). In these conditions, the collector with the slope angle θ=0° presents a turbine position near to the absorber. The difference appears in the maximum value, which is equal to V=2.41 m.s^{-1} with the slope angle θ=0°, V=2.01 m.s^{-1} with the slope angle θ=5°, V=2.06 m.s^{-1} with the slope angle θ=10° and V=2.14 m.s^{-1} with the slope angle θ=15°.

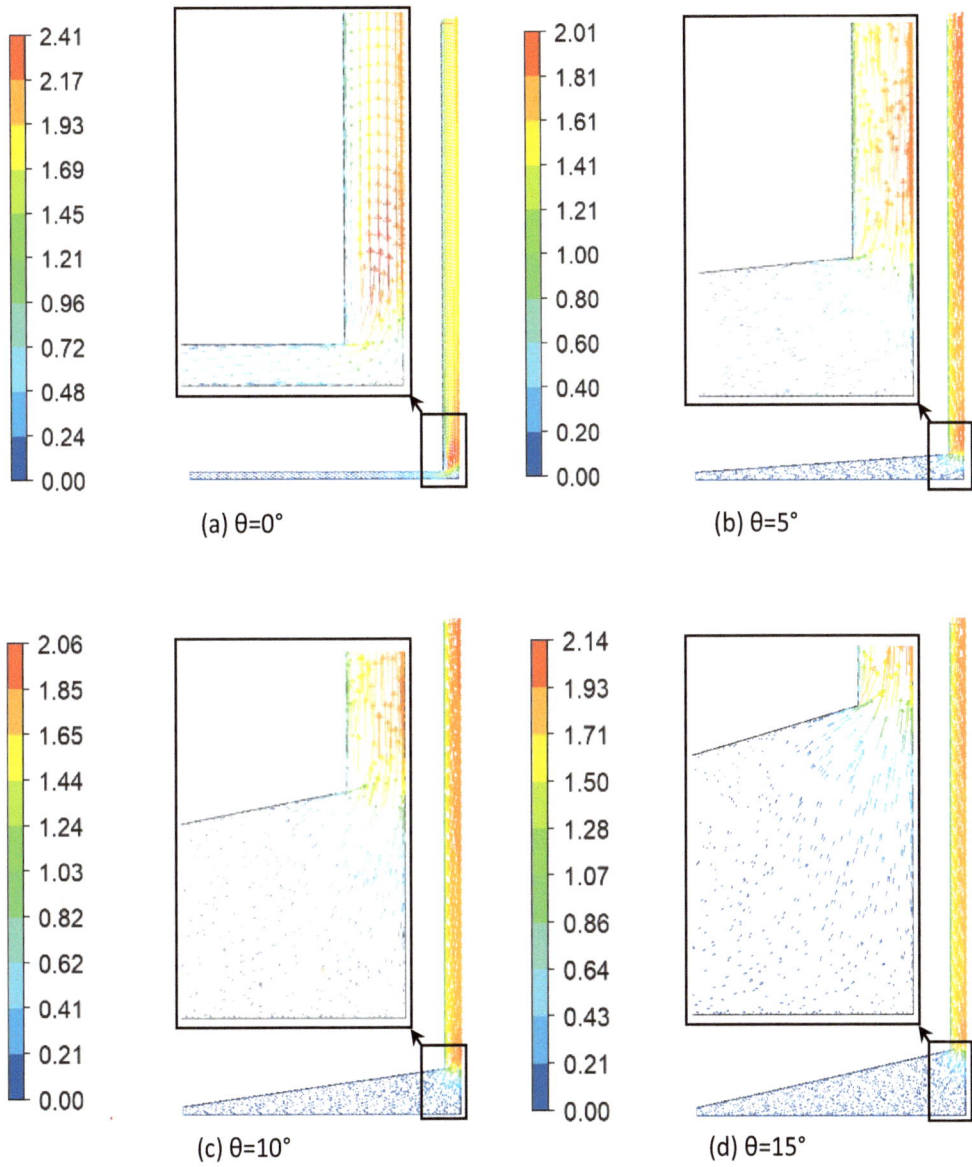

Fig. (4.22). Velocity fields for different collector slope angles.

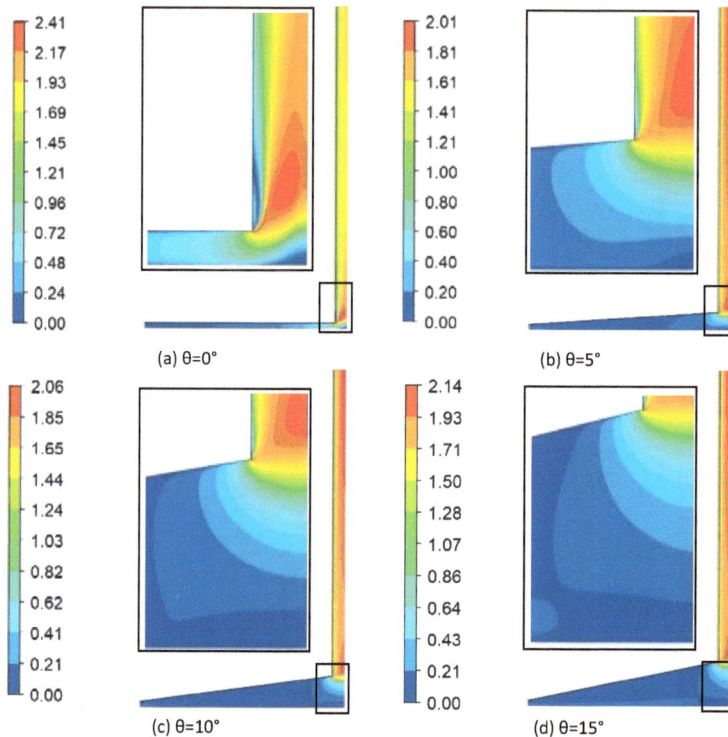

Fig. (4.23). Distribution of the magnitude velocity for different collector slope angles.

4.2.2. Temperature

Fig. (4.24) shows the profiles of the temperature in the collector in the axisymmetric plane of the solar chimney for different collector slope angles equal to $\theta=0°$, $\theta=5°$, $\theta=10°$, and $\theta=15°$. According to these results, it is clear that the variation of the collector slope angle does not have a great effect on the temperature distribution along with the collector. The temperature is equal to the ambient temperature in the collector input and it increases due to the greenhouse effect to reach its maximum at the collector output. Indeed, it has been observed that the temperature for the angle $\theta=0°$ is slightly higher than the other slope angles since in this case the volume of the system is low compared to the other cases. Fig. (4.25) shows the distribution of the temperature in the axisymmetric plane of the solar chimney for different collector slope angles. According to these results, the common point between the different cases confirms that the maximum value is located at the absorber since it has a higher coefficient of absorption the higher.

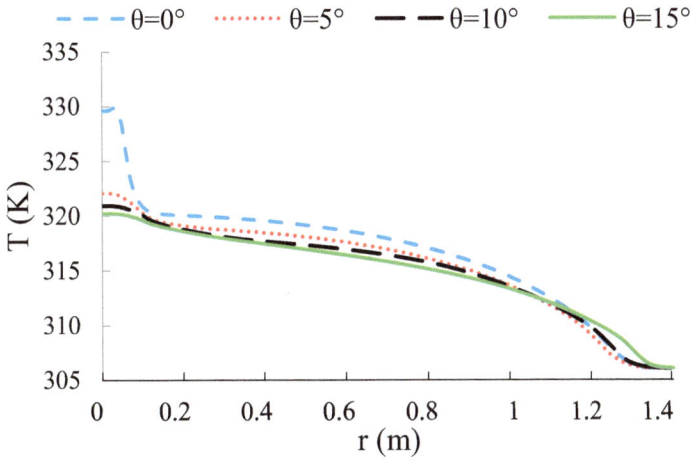

Fig. (4.24). Profiles of the temperature in the collector.

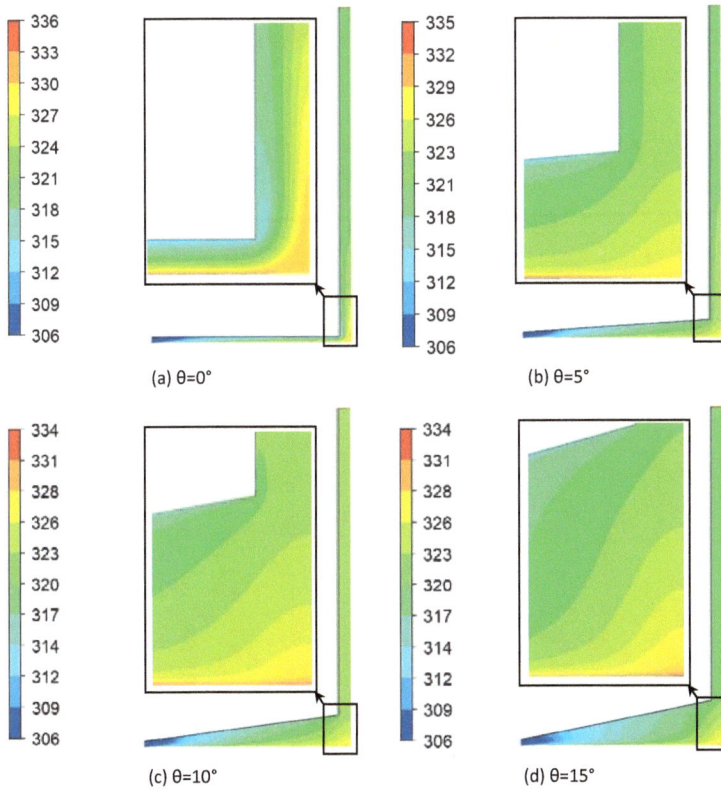

(a) θ=0°

(b) θ=5°

(c) θ=10°

(d) θ=15°

Fig. (4.25). Distribution of the temperature for different collector slope angles.

4.2.3. Static Pressure

Fig. (**4.26**) shows the static pressure profiles along the chimney axis for different collector slope angles equal to $\theta=0°$, $\theta=5°$, $\theta=10°$, and $\theta=15°$. According to these results, it is clear that the static pressure presents the same distribution for all collector slope angles. The depression zone characteristic of the maximum value is localized at the chimney inlet where the velocity is maximum. Then, the static pressure increases and reaches the atmospheric pressure value at the chimney outlet. Otherwise, this pressure difference is responsible for the airflow along with the chimney. Besides, it has been observed that the increase of the collector slope angle leads to decrease the depression at the base of the chimney. This fact is illustrated from the static pressure distribution in the axisymmetric plane of the chimney axis for different collector slope angles (Fig. **4.27**). According to these results, it is clear that the depression zone is located in the base of the chimney. Elsewhere, the static pressure increases until reaching the atmospheric pressure in the inlet of the collector and the outlet of the chimney. In these conditions, the difference is clear at the maximum value equal to p=4.47 Pa with $\theta=0°$, to p=4.31 Pa with $\theta=5°$, to p=4.35 Pa with $\theta=10°$ and to p=4.34 Pa with $\theta=15°$.

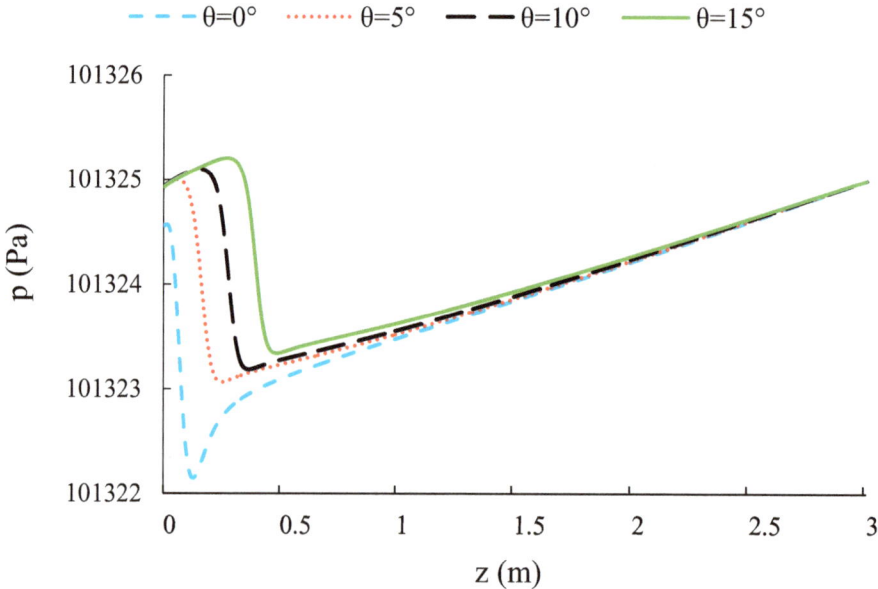

Fig. (4.26). Profiles of the static pressure along the chimney axis.

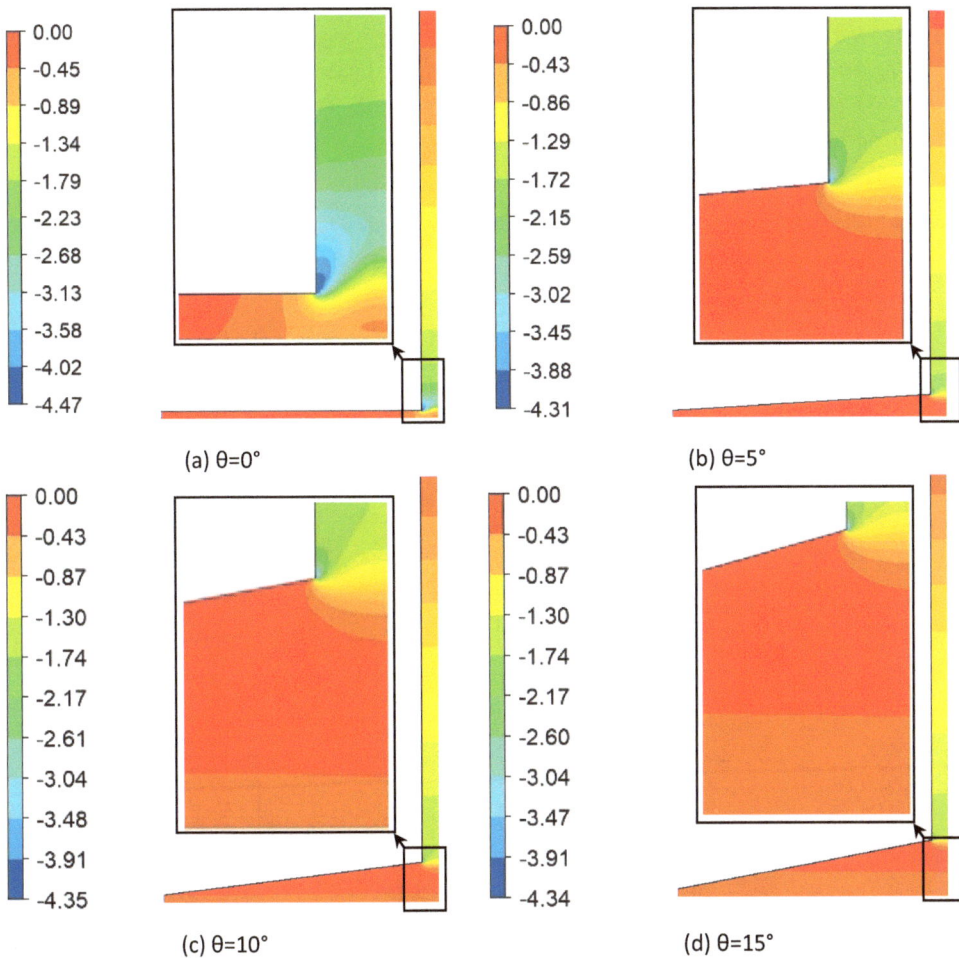

Fig. (4.27). Distribution of the static pressure for different collector slope angles.

4.2.4. Turbulent Kinetic Energy

Fig. (**4.28**) shows the distribution of the turbulent kinetic energy in the defined plane of the solar chimney for the considered collector slopes angles equal to θ= 0°, θ= 5°, θ= 10° and θ=15°. Based on these results, the turbulent kinetic energy gives an identical distribution for all slopes angles. Indeed, it is clear that the turbulent kinetic energy is found to be very weak in the axisymmetric plane excepting in the chimney entrance because the magnitude velocity in this zone is extreme. Likewise, it seems that there is a minor difference between the distributions of the turbulent kinetic energy for each case. This difference is located in the chimney inlet when the highest value is found in the third case with the collector slope angle θ= 10°. The maximum value of the turbulent kinetic energy is equal to k=0.367 $m^2.s^{-2}$ with slope angle θ=0°, k=0.566 $m^2.s^{-2}$ with the slope angle θ=5°, to k=0.602 $m^2.s^{-2}$ with slope angle θ=10°, to k=0.598 $m^2.s^{-2}$ with slope angle θ=15°.

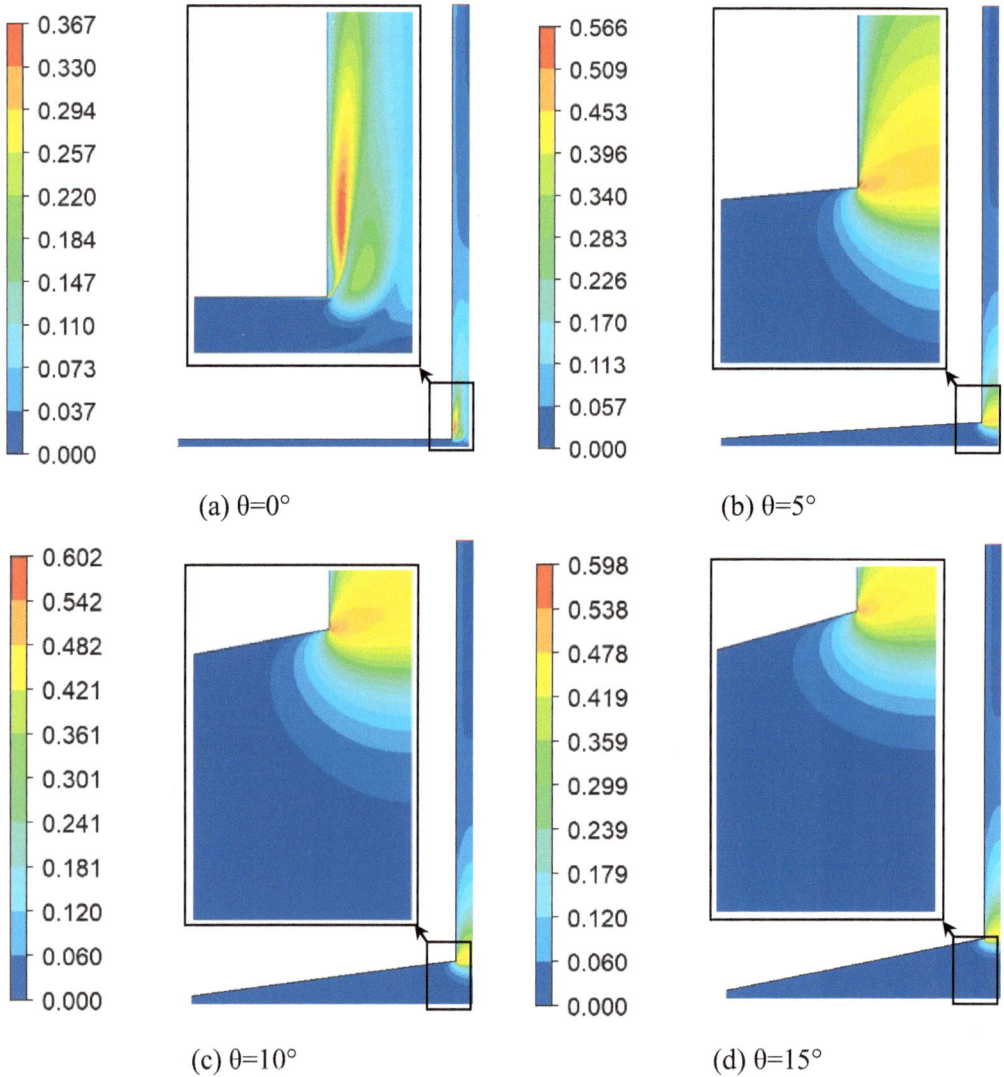

(a) θ=0°   (b) θ=5°

(c) θ=10°   (d) θ=15°

Fig. (4.28). Distribution of the turbulent kinetic energy for different collector slope angles.

4.2.5. Dissipation Rate of the Turbulent Kinetic Energy

Fig. (**4.29**) displays the distribution of the dissipation rate of the turbulent kinetic energy in the defined plane of the solar chimney for the different collector slope angles equal to θ=0°, θ=5°, θ=10°, and θ=15°. From these results, the variation of the collector slope angle does not affect the distribution of the dissipation rate of the turbulent kinetic energy. It is found to be very weak in the axisymmetric plane excluding in the collector exit. Certainly, it is clear that the dissipation rate of the turbulent kinetic energy growths with the growth of the collector slope angle. The comparison of these results, confirms that the collector slop angle has a direct effect on the maximum value. Which is equal to ε=368 $m^2.s^{-3}$ with collector slope angle θ=0°, to ε=367 $m^2.s^{-3}$ with collector slope angle θ=5°, to ε=444 $m^2.s^{-3}$ with collector slope angle θ=10° and to ε=452 $m^2.s^{-3}$ with collector slope angle θ=15°.

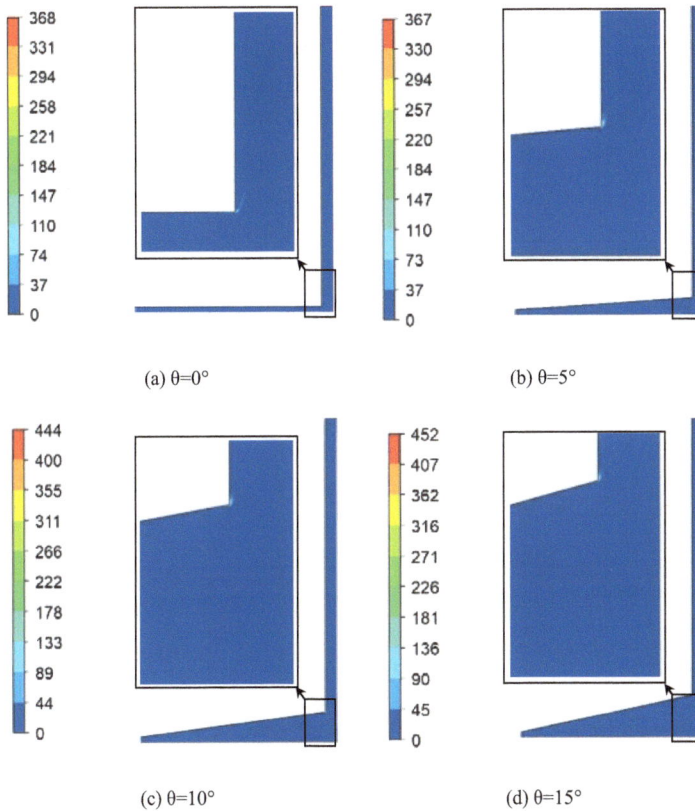

(a) θ=0°

(b) θ=5°

(c) θ=10°

(d) θ=15°

Fig. (4.29). Distribution of the dissipation rate of the turbulent kinetic energy for different collector slope angles.

4.2.6. Turbulent Viscosity

Fig. (**4.30**) shows the distribution of the turbulent viscosity in the axisymmetric plane of the solar chimney for the different collector slopes angles equal to $\theta=0°$, $\theta=5°$, $q=10°$, and $\theta=15°$. According to these results, it is clear that the turbulent viscosity is found to be very weak in the axisymmetric plane except in the chimney inlet because the magnitude velocity in this place is maximum. Indeed, it appears that there is a small difference between the distribution of the turbulent viscosity for each slope angle. This difference is located in the chimney inlet when the highest value is found in the third case with the collector slope angle $\theta=10°$. The minimum value of the turbulent viscosity is equal to $\mu_t =0.0026$ kg.m^{-1}.s^{-1} with a slope angle of $\theta=0°$. However, the maximum value for the turbulent viscosity is obtained with the slope angle $\theta=10°$ and it is equal to $\mu_t =0.00131$ kg.m^{-1}.s^{-1}.

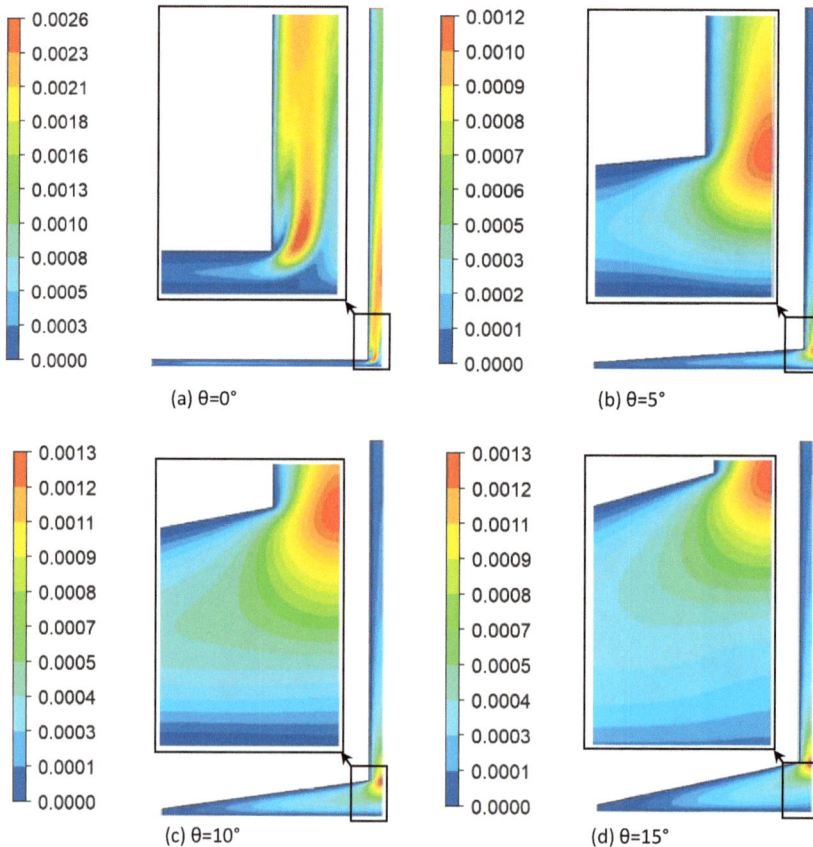

Fig. (**4.30**). Distribution of the turbulent viscosity for different collector slope angles.

4.3. Collector Height Effect

In this section, we are interested in the study of the collector height effect on the local results such as the magnitude velocity, the temperature, the static pressure, and the turbulence characteristics. The system considered is defined by a collector slope angle equal to $\theta=0°$, a chimney diameter d= 0.16 m, a collector diameter D= 2.75 m, and a chimney height H=3 m (Fig. **4.31**). Particularly, we have studied the solar chimney for four collector height equal to h=0.005 m, h=0.05 m, h=0.075 m, and h=0.1 m. The different considered configurations are shown in Fig. (**4.32**).

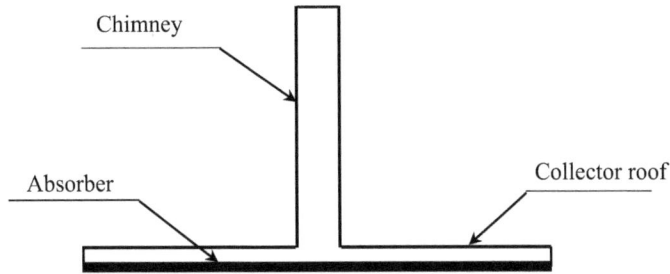

Fig. (4.31). Schematic of the system.

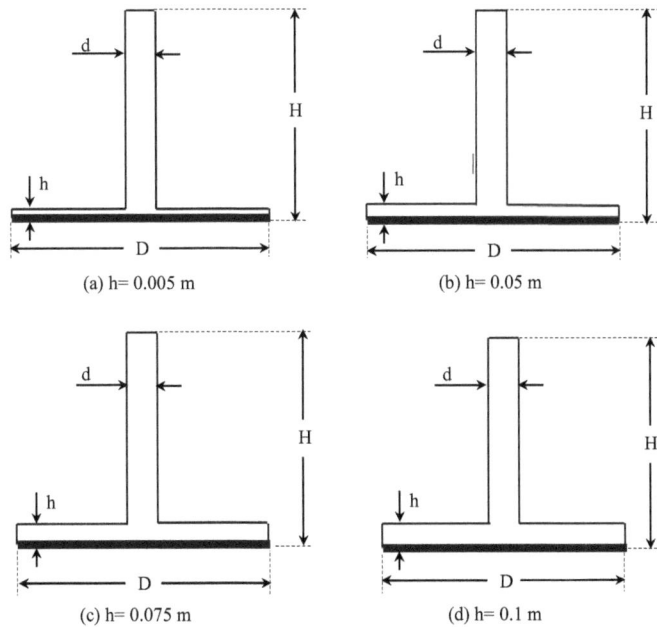

(a) h= 0.005 m   (b) h= 0.05 m

(c) h= 0.075 m   (d) h= 0.1 m

Fig. (4.32). Schematic of the different geometries.

4.3.1. Magnitude Velocity

Fig. (**4.33**) shows the variation of the maximum magnitude velocity for different collector heights. According to these results, it is clear that the velocity has an optimum value for the collector height equal to h = 0.05 m. Indeed, the maximum value is reduced if the collector height increases due to the reduction of the air heat transfer. However, with the small collector height h=0.005 m the magnitude velocity is low because the air volume is low and not able to create an airflow. Fig. (**4.34**) shows the profile of the magnitude velocity along the chimney axis. According to these results, it is clear that the magnitude velocity remains the same distribution for all configurations. It is equal to V= 0 m.s^{-1} at the absorber (z=0 m) and it increases until reaching the maximum value at the chimney inlet and it remains uniform until the chimney outlet with a light decrease.

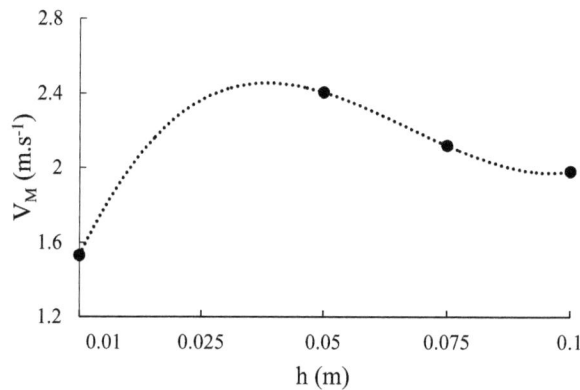

Fig. (4.33). Variation of the maximum magnitude velocity.

The distribution of the velocity fields and the magnitude velocity in the axisymmetric plane of the solar chimney for different collector heights equal to h=0.005 m, h=0.05 m, h=0.075 m and h=0.1 m is shown respectively in Figs. (**4.35**) and (**4.36**). According to these results, it is clear that the magnitude velocity is low at the collector and the acceleration zone is localized at the chimney inlet. Moreover, the velocity increases at the chimney inlet except in the first case defined by the height h=0.005 m, where the maximum value of the magnitude velocity appears in the collector outlet. Therefore, on a large scale, there is some station where the turbines are placed horizontally around the base of the chimney. By comparing the results for the different height, it is clear that the difference is obtained at the maximum value. In these conditions, the maximum value of the magnitude velocity is equal to V=1.54 m.s^{-1} with h=0.005 m, to V=2.41 m.s^{-1} with h=0.05 m, to V=2.12 m.s^{-1} with h=0.075 m and to V=2.06 m.s^{-1} with h= 0.1 m.

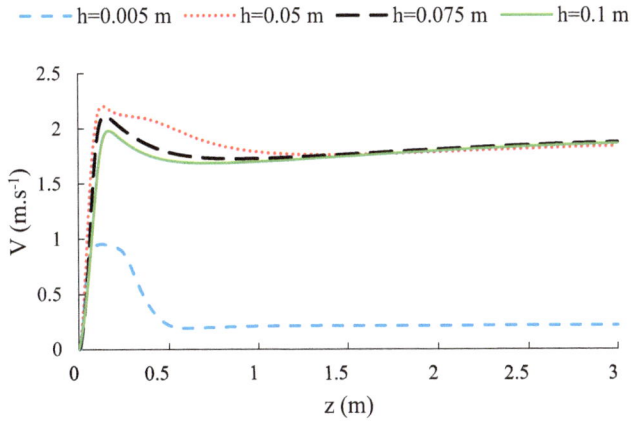

Fig. (4.34). Profiles of the magnitude velocity along the chimney axis.

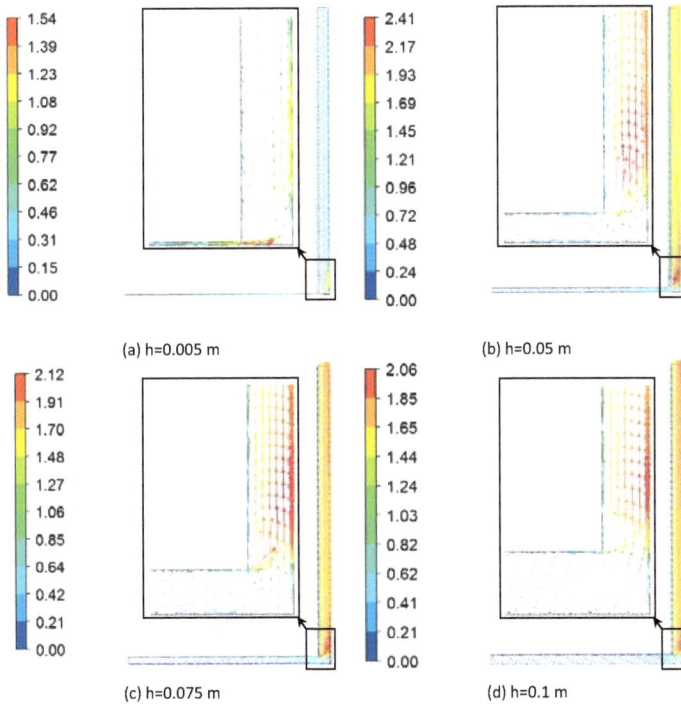

(a) h=0.005 m

(b) h=0.05 m

(c) h=0.075 m

(d) h=0.1 m

Fig. (4.35). Velocity fields for different collector heights.

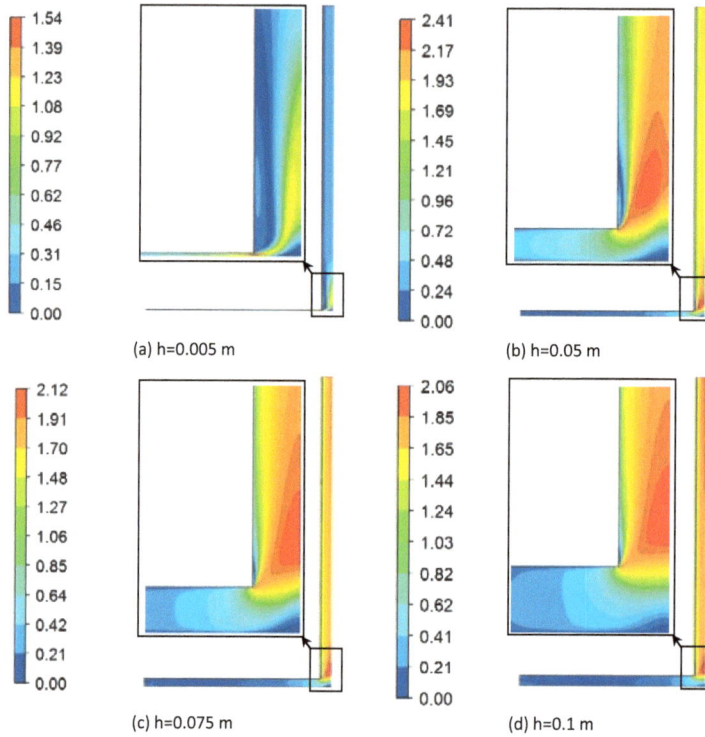

Fig. (4.36). Distribution of the magnitude velocity for different collector heights.

4.3.2. Temperature

Fig. (**4.37**) shows the temperature profiles in the collector. According to these results, it is clear that the temperature for the first case defined by h=0.005 m is higher than the other cases because the air volume is very low, so the heat transfer rate presents the highest value. Indeed, the wind has a lower effect on the airflow if the collector inlet is low. It has been observed that the temperature decreases with the increase of the collector height due to the increase of the air volume. Fig. (**4.38**) presents the distribution of the temperature in the axisymmetric plane of the solar chimney. According to these results, it is clear that the temperature presents the same distribution for the different collector heights. Indeed, it has been observed that the maximum value of the temperature is concentrated in the absorber and is degraded until the ambient temperature at the collector roof. Besides, the variation of the collector height doesn't affect the maximum temperature value.

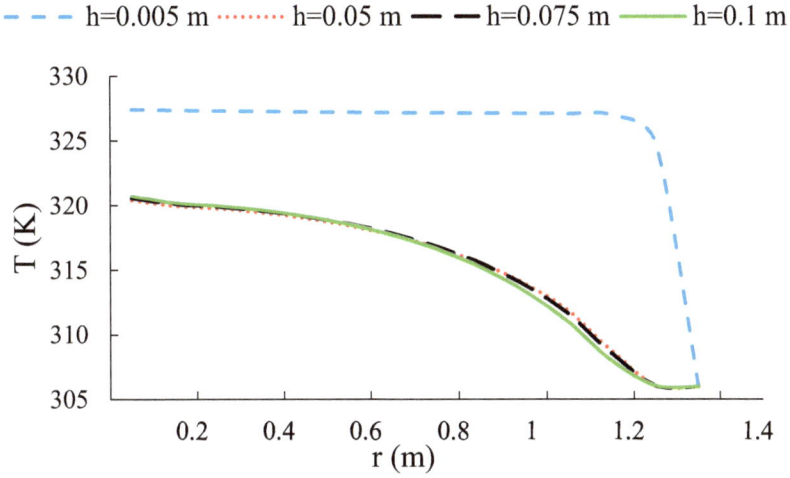

Fig. (4.37). Profiles of the temperature in the collector.

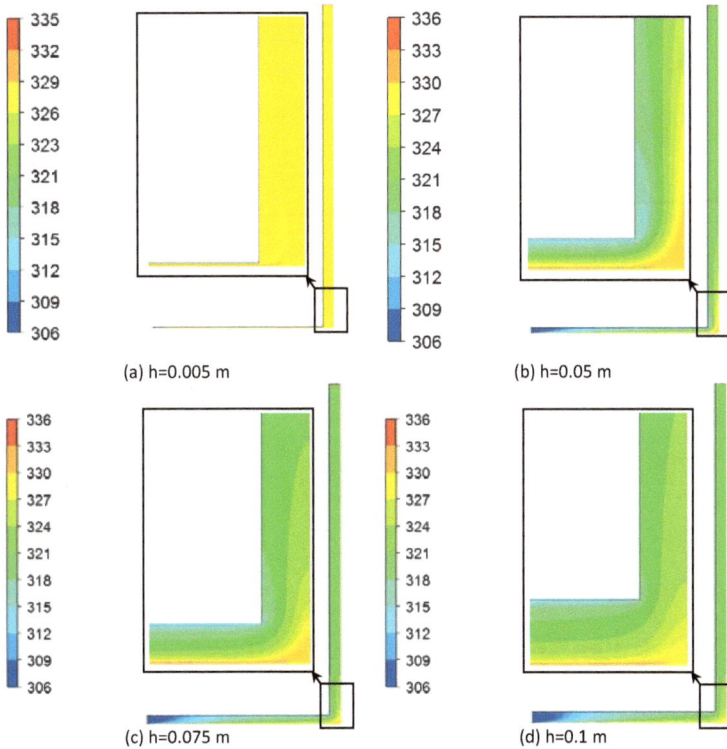

Fig. (4.38). Distribution of the temperature for different collector heights.

4.3.3. Static Pressure

Fig. (**4.39**) shows the static pressure profiles along the chimney axis. According to these results, the static pressure presents the same distribution for the different collector heights. The depression zone is located in the chimney base. Along with the chimney, the pressure increases and reaches the atmospheric pressure in the outlet. Besides, Fig. (**4.40**) shows the distribution of the static pressure. According to these results, it is clear that the pressure zone is located in the collector and at the chimney outlet while the depression zone is located in the base of the chimney. The comparison between results confirms that the maximum value is equal to p=4.56 Pa with a high h=0.005 m. However, the minimum value is equal to p=4.22 Pa with a high h=0.075 m.

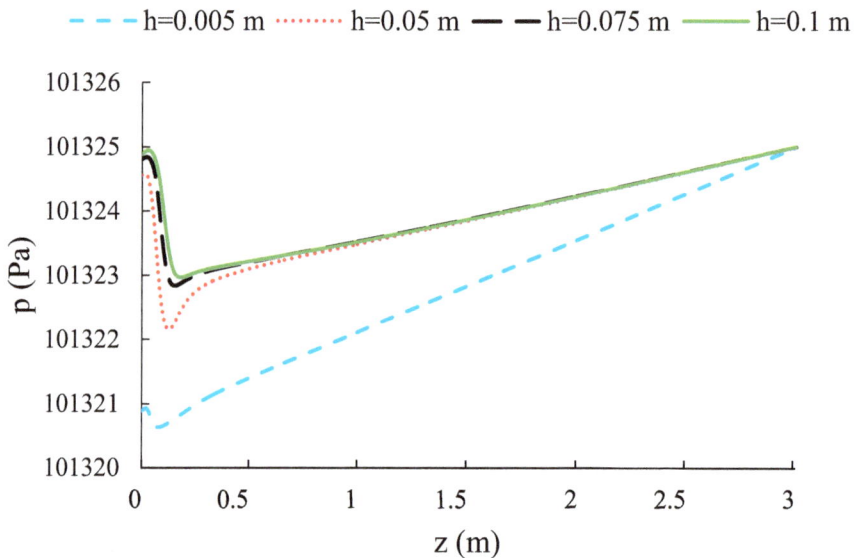

Fig. (4.39). Profiles of the static pressure along the chimney axis.

4.3.4. Turbulent Kinetic Energy

Fig. (**4.41**) shows the distribution of the turbulent kinetic energy in the axisymmetric plane of the solar chimney for the different collector heights equal to h= 0.005 m, h= 0.05 m, h= 0.075 m and h= 0.1 m. According to these results, the turbulent kinetic energy presents the same distribution for all heights. Indeed, it is clear that the turbulent kinetic energy is found to be very weak in the axisymmetric plane except in the chimney inlet since the magnitude velocity in this zone is maximum. The difference between the four cases is located in the chimney inlet, where the highest value is found. It has been observed that the turbulent kinetic energy increases with the increase of the collector height. The maximum value is equal to k=0.160 $m^2.s^{-2}$ with the height h=0.005 m, to k=0.367 $m^2.s^{-2}$ with the height h=0.05 m, to k=0.377 $m^2.s^{-2}$ with the height h=0.075 and k=0.458 $m^2.s^{-2}$ with the height h=0.1 m.

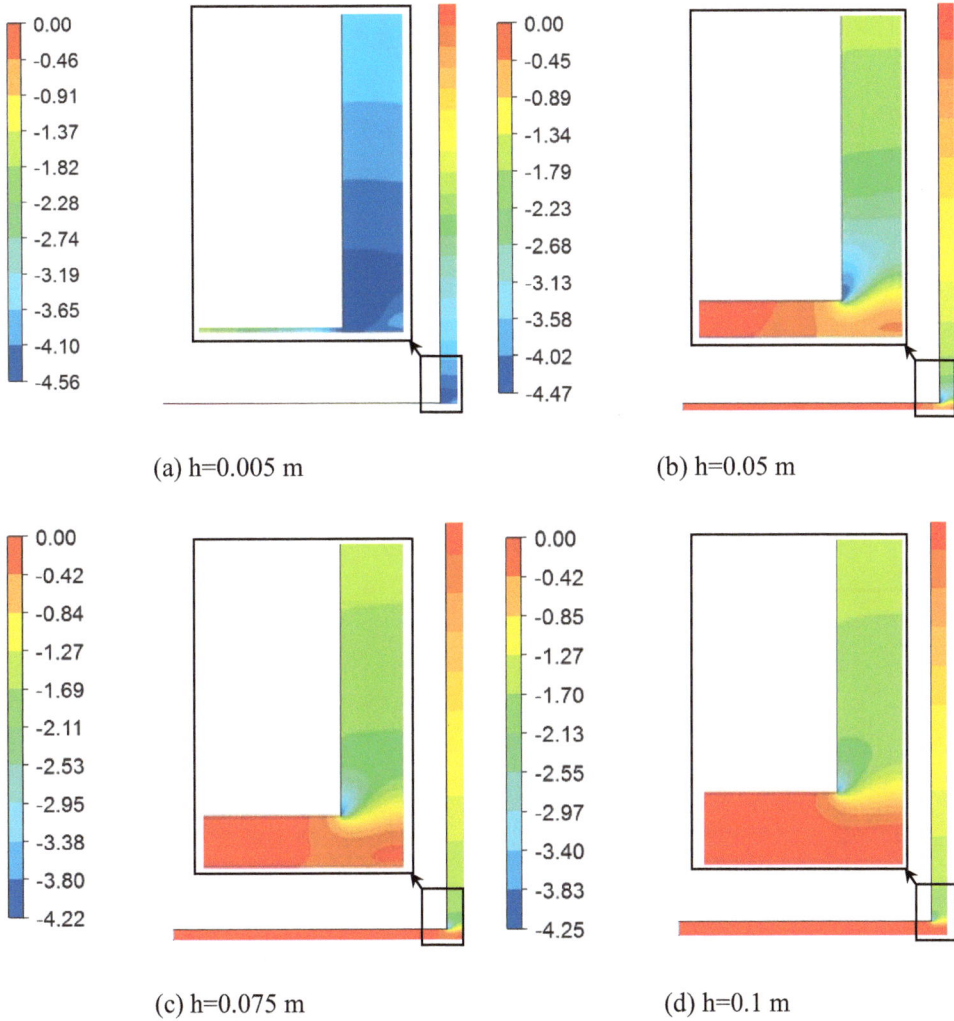

(a) h=0.005 m   (b) h=0.05 m

(c) h=0.075 m   (d) h=0.1 m

Fig. (4.40). Distribution of the static pressure for different collector heights.

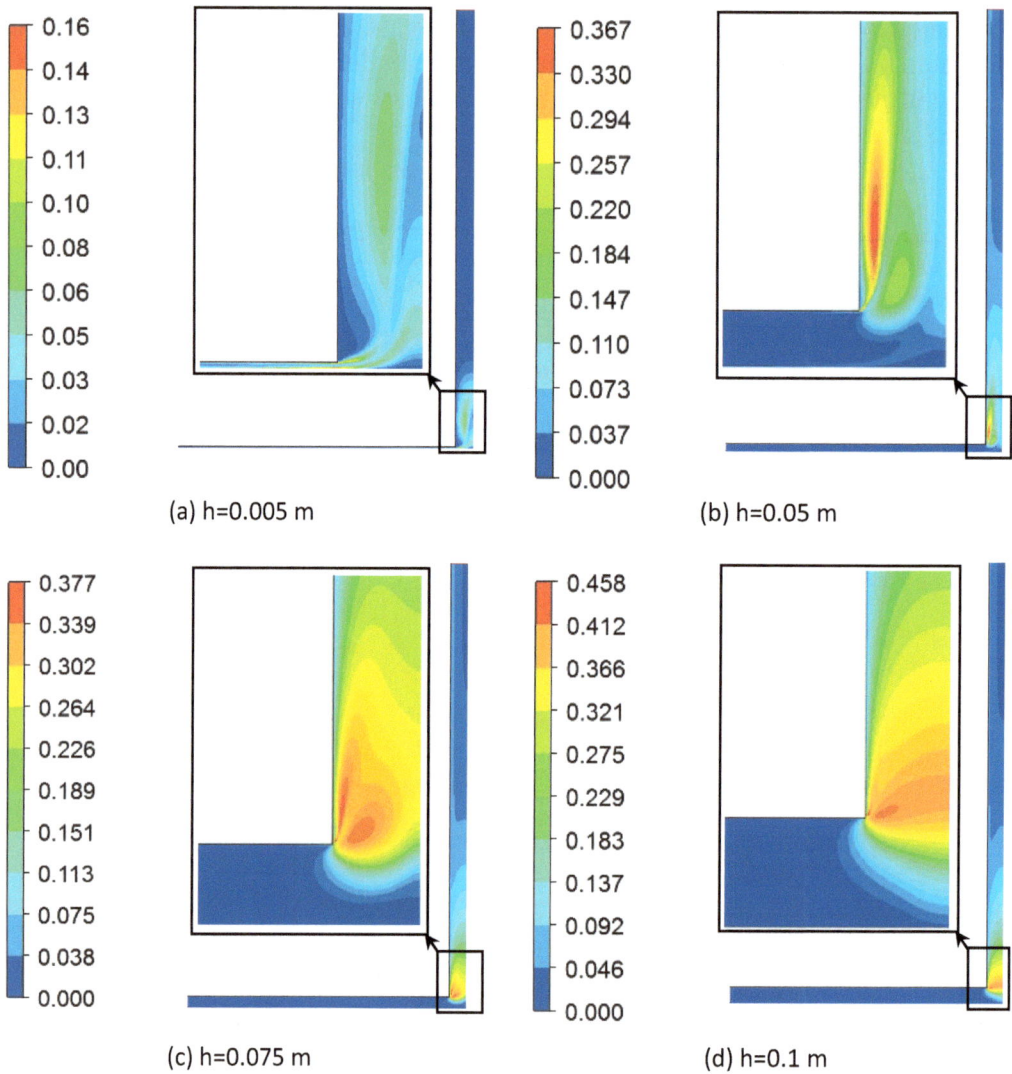

Fig. (4.41). Distribution of the turbulent kinetic energy for different collector heights.

4.3.5. Dissipation Rate of the Turbulent Kinetic Energy

Fig. (**4.42**) shows the distribution of the dissipation rate of the turbulent kinetic energy in the axisymmetric plane of the solar chimney for the different collector heights equal to h=0.005 m, h=0.05 m, h=0.075 m and h=0.1 m. According to these results, the dissipation rate of the turbulent kinetic energy presents the same distribution for all collector heights. Indeed, it is found to be very weak in the axisymmetric plane except in the collector outlet. This fact is due to the important value of the viscous shear stress that appeared on the airflow in the solar chimney. Besides, the dissipation rate of the turbulent kinetic energy reaches the maximum value with h=0.05 m. In fact, in this case the magnitude velocity is still maximum. The difference between results appears in the maximum value of the dissipation rate of the turbulent kinetic energy.

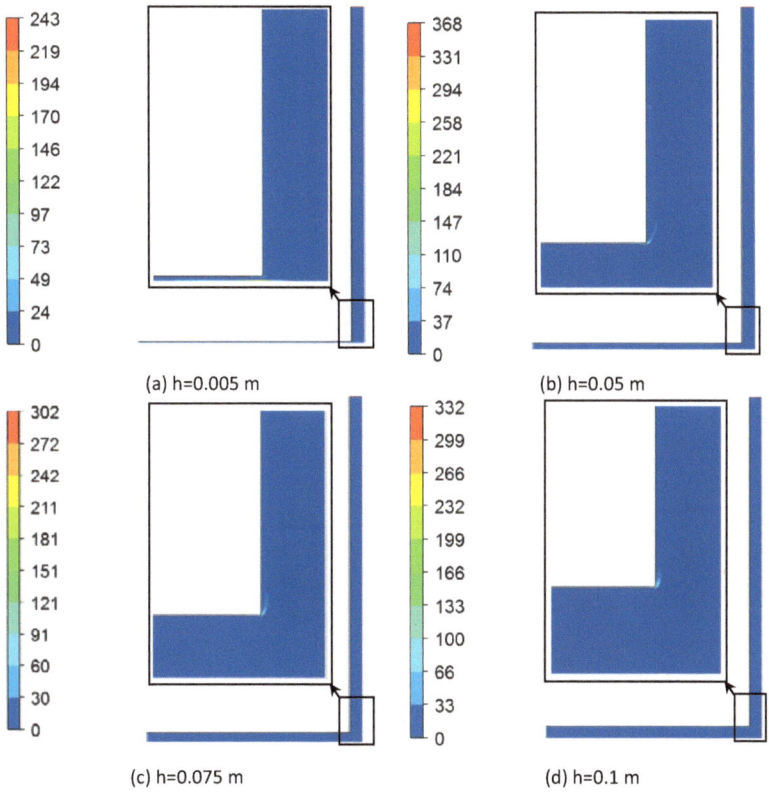

Fig. (4.42). Distribution of the Dissipation rate of the turbulent kinetic energy for different collector heights.

4.3.6. Turbulent Viscosity

Fig. (**4.43**) shows the distribution of the turbulent viscosity in the axisymmetric plane of the solar chimney for the different collector heights equal to h= 0.005 m, h= 0.05 m, h= 0.075 m and h= 0.1 m. According to these results, the turbulent viscosity presents the same distribution for all heights. The turbulent viscosity is found to be very weak in the axisymmetric plane except in the chimney inlet since the magnitude velocity in this zone is maximum. The difference between the four cases is clear in the chimney inlet when the highest value is found. It has been observed that the turbulent viscosity increases with the increase of the collector height. In these condition, the maximum value is equal to $\mu_t = 0.00154$ kg.m^{-1}.s^{-1} with the height h= 0.005 m, to $\mu_t = 0.0026$ kg.m^{-1}.s^{-1} with the height h= 0.05 m, to $\mu_t = 0.00543$ kg.m^{-1}.s^{-1} with the height h= 0.075 and to $\mu_t = 0.00828$ kg.m^{-1}.s^{-1} with the height h= 0.1 m.

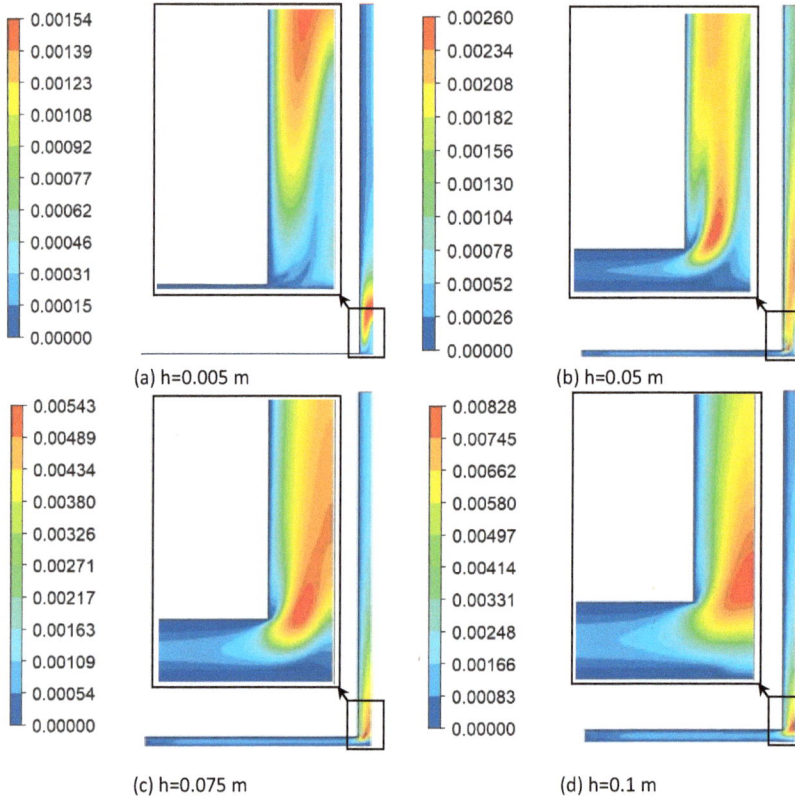

(a) h=0.005 m

(b) h=0.05 m

(c) h=0.075 m

(d) h=0.1 m

Fig. (4.43). Distribution of the turbulent viscosity for different collector heights.

4.4. Chimney Diameter Effect

In this section, we are interested in the study of the chimney diameter effect on the local results such as the magnitude velocity, the temperature, and the static pressure and the turbulence characteristic. The system considered is defined by a collector slope angle equal to $\theta=0°$, a collector height h= 0.005 m, a collector diameter D= 2.75 m, and a chimney height H=3 m (Fig. **4.44**). Particularly, we have studied the solar chimney for four chimney diameter equal to d=0.1 m, d=0.13 m, d=0.16 m, and d=0.2 m. The different geometries are shown in Fig. (**4.45**).

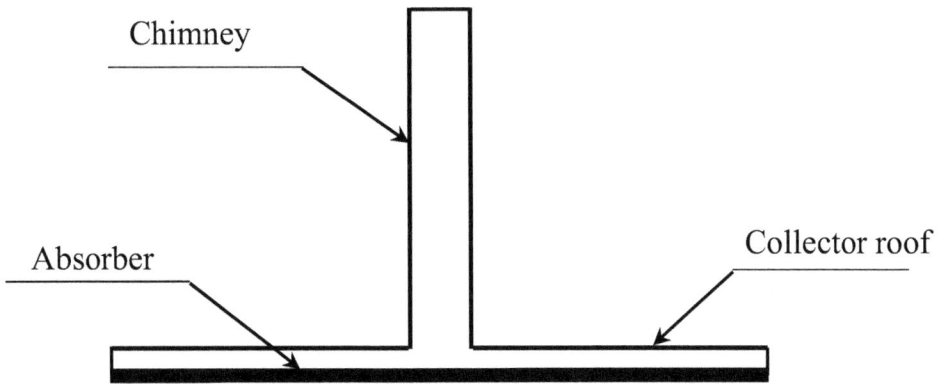

Fig. (4.44). Schematic of the system.

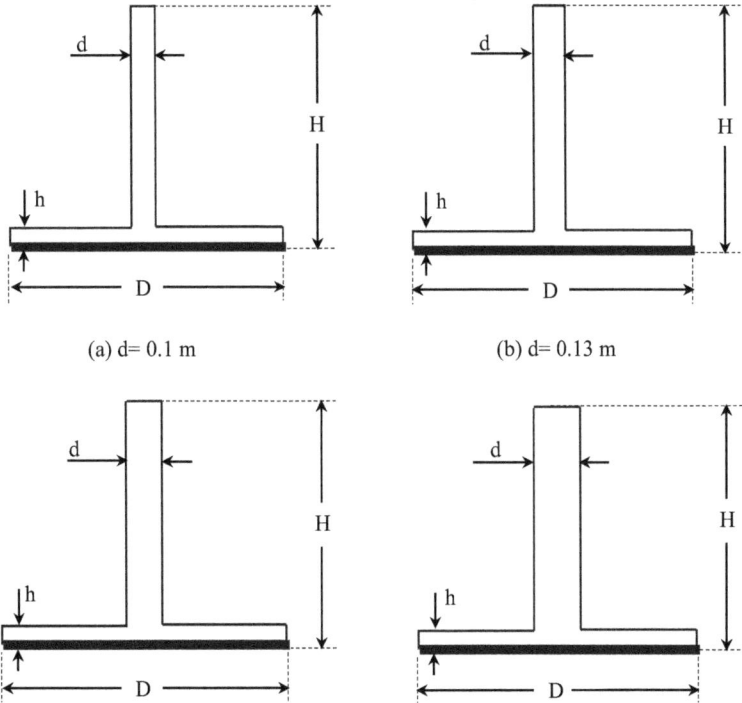

(a) d= 0.1 m   (b) d= 0.13 m

Fig. (4.45). Schematic of the different geometries.

4.4.1. Magnitude Velocity

The chimney diameter is one of the most important parameters in the solar chimney power plant. Fig. (**4.46**) shows the variation of the maximum magnitude velocity value for the different chimney diameters. According to these results, it is clear that the magnitude velocity in the system increases with the increase of the chimney diameter. Consequently, a greater volume of air will be heated in the collector for the same period. Thereby, the maximum velocity values decrease with the increase of the tower diameter. Indeed, the maximum diameter equal to d=0.2 m gives the best efficiency for the system. Figs. (**4.47**) & (**4.48**) show the profiles of the magnitude velocity along with the collector for different chimney diameter.

According to these results, it has been observed that the velocity along with the collector increase with the increase of the chimney diameter.

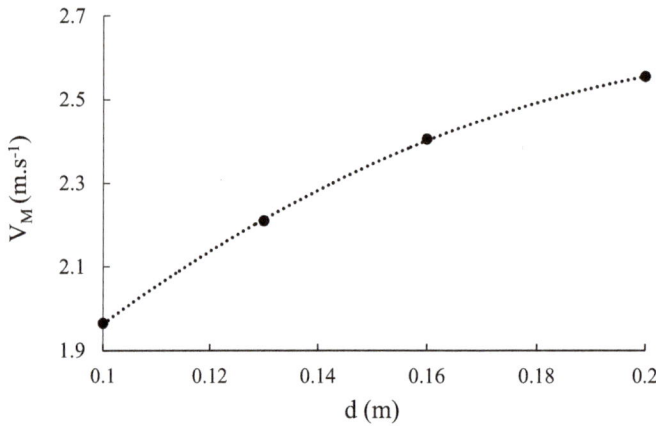

Fig. (4.46). Variation of the maximum magnitude velocity.

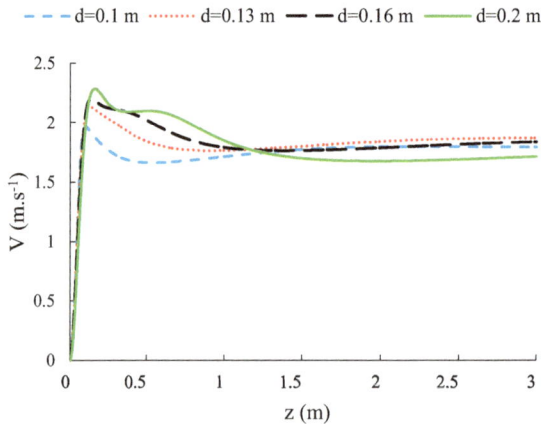

Fig. (4.47). Profiles of the magnitude velocity along the chimney axis.

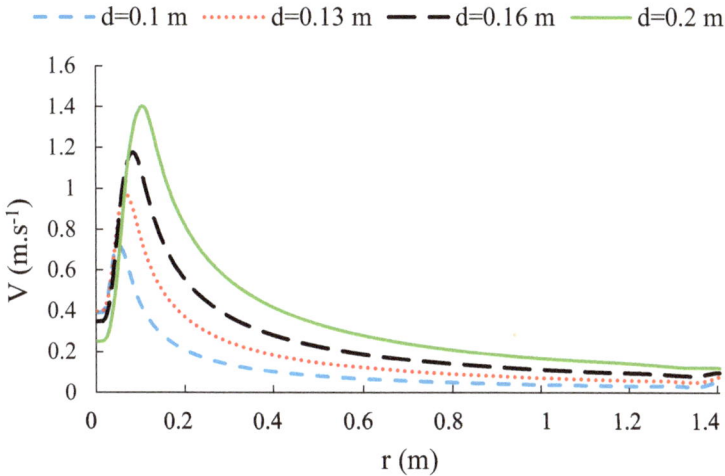

Fig. (4.48). Profiles of the magnitude velocity in the collector.

The distribution of the velocity fields and the magnitude velocity in the axisymmetric plane of the solar chimney for different chimney diameters equal to d=0.1 m, d=0.13 m, d=0.16 m, and d=0.2 m is shown respectively in Figs. (**4.49**) and (**4.50**). According to these results, it is clear that the magnitude velocity presents the same distribution for the different chimney diameter. Indeed, it is illustrated that the maximum magnitude velocity is concentrated in the chimney inlet. It has been observed that the increase in the chimney diameter produces an increase in the magnitude velocity. Therefore, a greater volume of air will be heated in the collector for the same period, thereby decreasing the maximum magnitude velocity values in the chimney. The difference between results is clear at the maximum value of the magnitude velocity which it is equal to V=1.97 m.s^{-1} with d=0.1 m, to V=2.24 m.s^{-1} with d=0.13 m, to V=2.41 m.s^{-1} with d=0.16 m and to V=2.56 m.s^{-1} with d=0.2 m.

4.4.2. Temperature

Fig. (**4.51**) shows the temperature profiles in the collector for different chimney diameters equal to d=0.1 m, d=0.13 m, d=0.16 m, and d=0.2 m. According to these results, it has been observed that the temperature increases in the collector center due to the heat exchanges between the absorber and the airflow. Indeed, it has been noted that the temperature in the collector increase with the increase of the chimney diameter. A larger volume of air needs to be heated in the cover during the same period, causing a decrease in the temperature. It can be noted that the flow temperature in the tower is always greater than the ambient temperature outside the solar chimney. Fig. (**4.52**) shows the distribution of the temperature in the axisymmetric plane of the solar chimney. According to these results, it is clear that the temperature presents the same distribution for the different chimney diameters. Indeed, the maximum temperature value is concentrated in the absorber and the chimney centerline. Besides, the variation of chimney diameter has been noted a great effect on the maximum value of the temperature, which is equal to T=338 K with d=0.1 m and d=0.13 m, to T=336 K with d=0.16 m and to T=335 K with d=0.2 m.

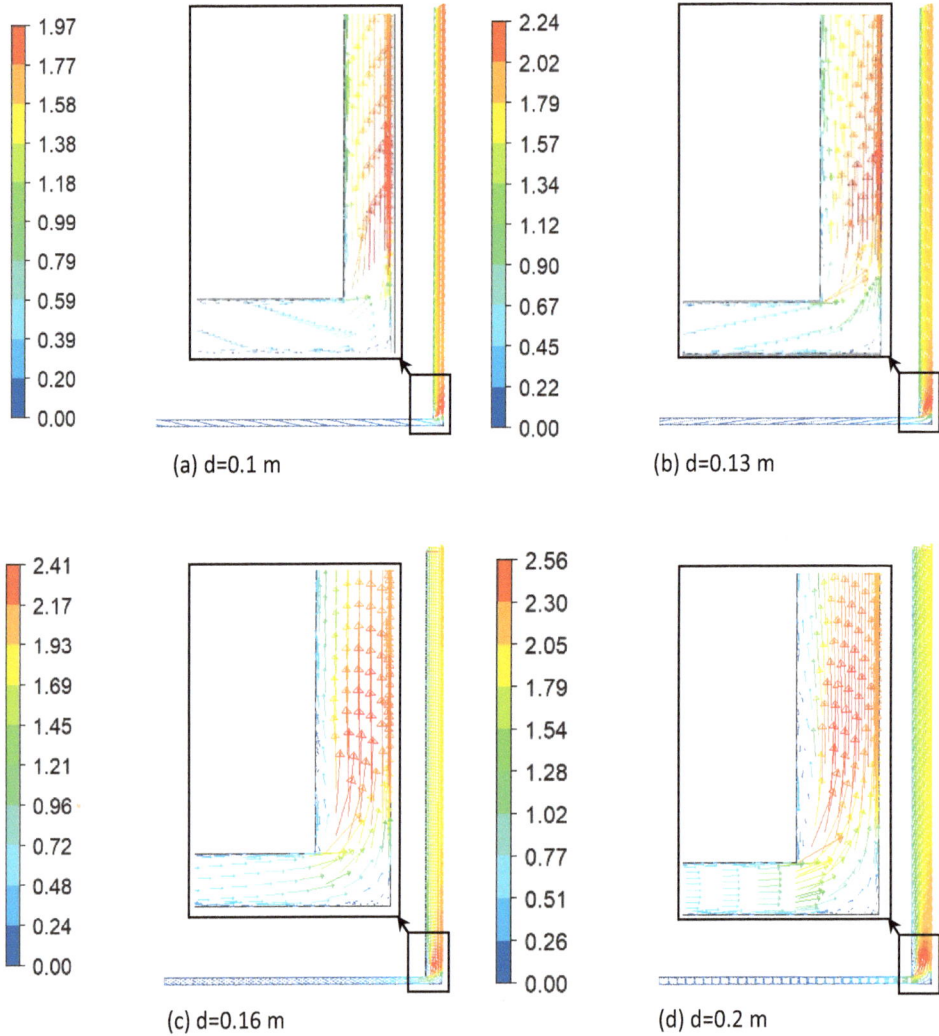

(a) d=0.1 m

(b) d=0.13 m

(c) d=0.16 m

(d) d=0.2 m

Fig. (4.49). Velocity fields for different chimney diameters.

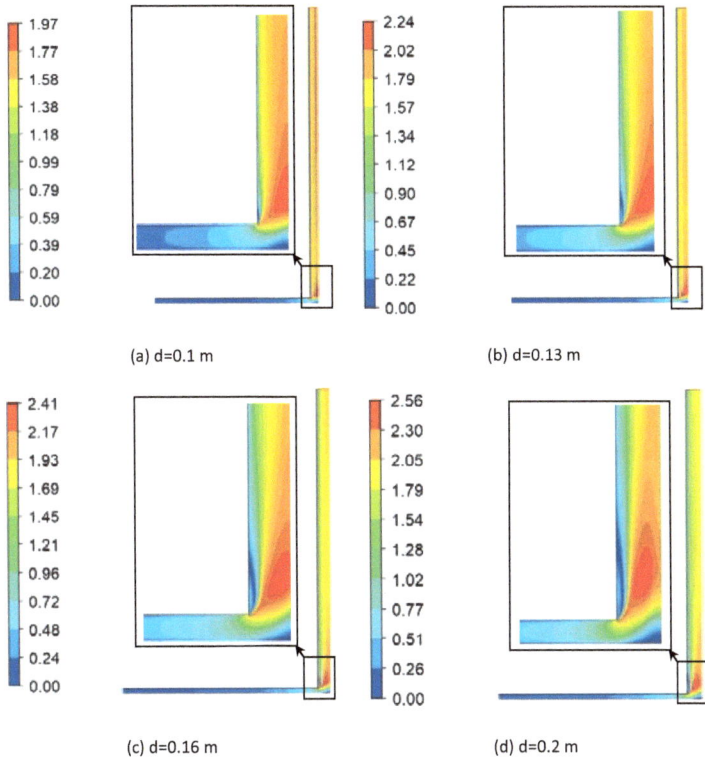

Fig. (4.50). Distribution of the magnitude velocity for different chimney diameters.

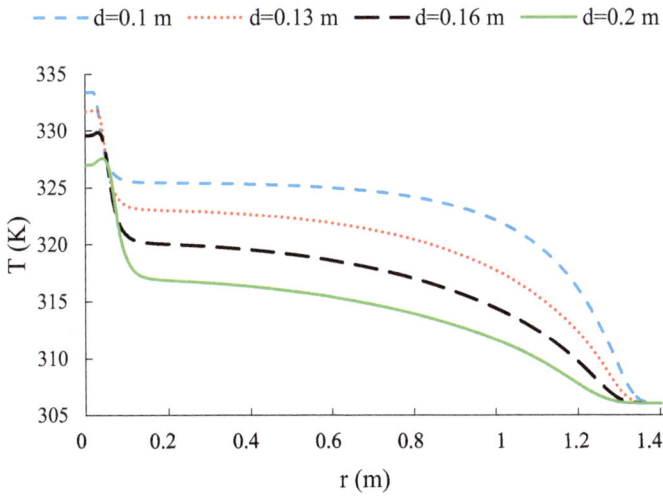

Fig. (4.51). Profiles of the temperature in the collector.

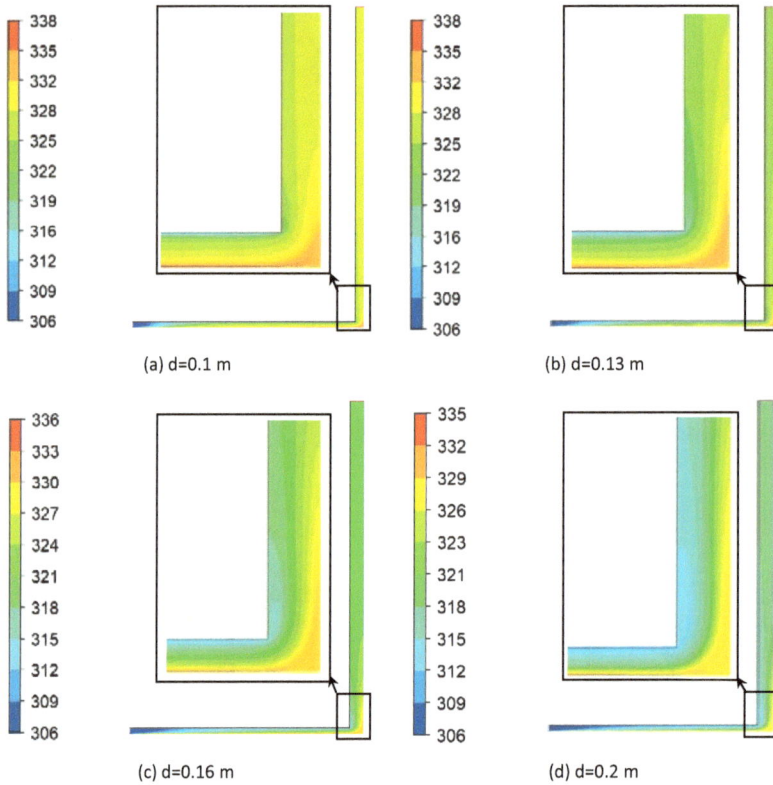

(a) d=0.1 m

(b) d=0.13 m

(c) d=0.16 m

(d) d=0.2 m

Fig. (4.52). Distribution of temperature for different chimney diameters.

4.4.3. Static Pressure

Fig. (**4.53**) shows the static pressure profiles along the chimney axis for different chimney diameters equal to d=0.1 m, d=0.13 m, d=0.16 m, and d=0.2 m. According to these results, it is clear that the static pressure presents the same distribution for all chimney diameter. The depression value is maximum at the chimney inlet where the velocity is maximum. Then, the static pressure increases and reaches the atmospheric pressure value at the chimney outlet. Otherwise, this pressure difference is responsible for the airflow along with the chimney. Besides, it has been observed that the increase of the chimney diameter leads to increase the depression at the base of the chimney. This fact is illustrated with the static pressure distribution in the axisymmetric plane of the chimney axis for different chimney diameter is presented in Fig. (**4.54**). According to these results, it is clear that the depression zone is located in the base of the chimney. Elsewhere, the static pressure increases until reaching the atmospheric pressure in the inlet of the collector and the outlet of the chimney. In these conditions, the difference is clear between the different configurations where the maximum value of the static pressure is equal to p=3.66 Pa with d=0.1 m, to p=4.20 Pa with d=0.13 m, to p=4.47 Pa with d= 0.16 m and to p=4.62 Pa with d=0.2 m.

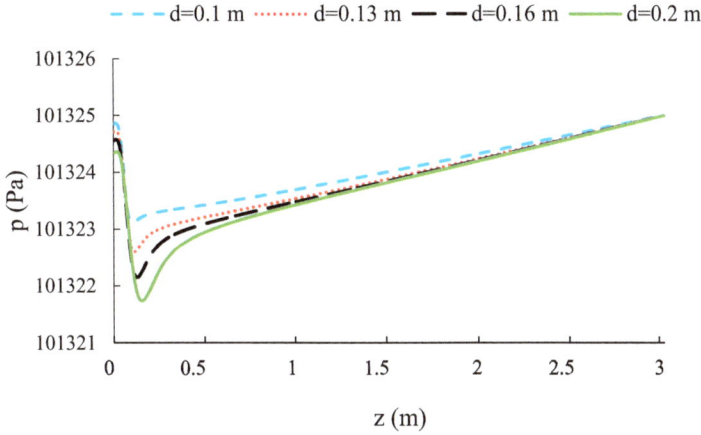

Fig. (4.53). Profiles of the static pressure along the chimney axis.

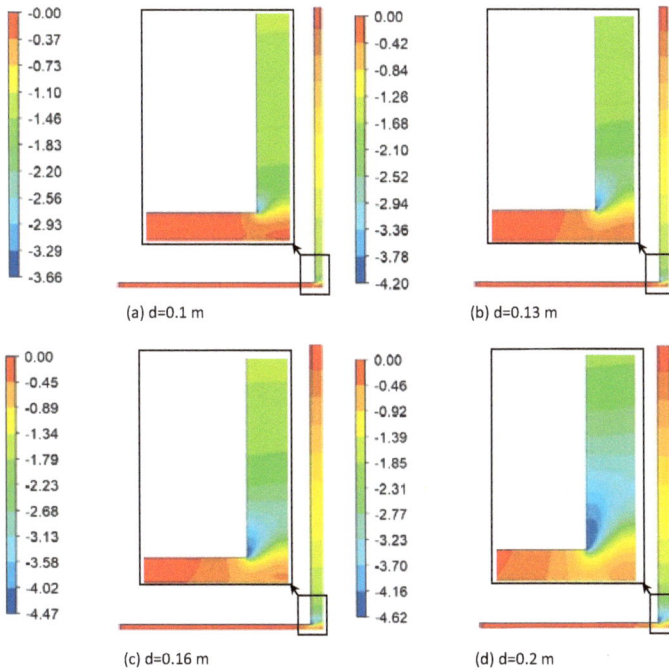

Fig. (4.54). Distribution of the static pressure for different chimney diameters.

4.4.4. Turbulent Kinetic Energy

Fig. (4.55) shows the distribution of the turbulent kinetic energy in the axisymmetric plane of the solar chimney for the different chimney diameters equal to d= 0.1 m, d= 0.13 m, d= 0.16 m, and d=0.2 m. According to these results, the turbulent kinetic energy presents the same distribution for all diameters. It is low in the axisymmetric plane except in the chimney inlet since the magnitude velocity in this zone is maximum. The difference between the four cases is located in the chimney inlet when the highest value is found. It has been observed that the turbulent kinetic energy increases with the increase of the chimney diameter. The maximum value of the turbulent kinetic energy is equal to k=0.354 m2.s-2 with the diameter d= 0.1 m, to k=0.337 m2.s-2 with the diameter d= 0.13 m, to k=0.367 m2.s-2 with the diameter d= 0.16 and to k=0.382 m2.s-2 with the diameter d= 0.2 m.

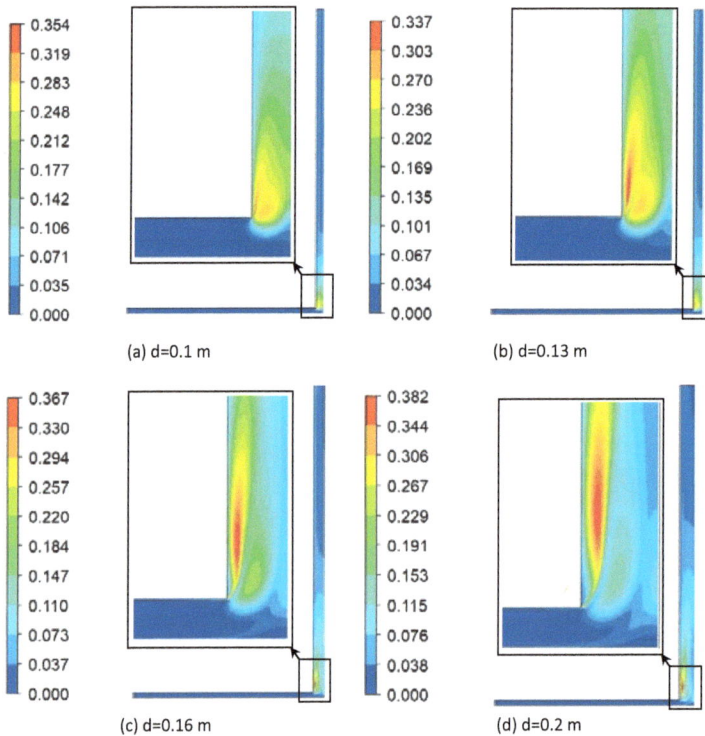

Fig. (4.55). Distribution of the turbulent kinetic energy for different chimney diameters.

4.4.5. Dissipation Rate of the Turbulent Kinetic Energy

Fig. (**4.56**) shows the distribution of the dissipation rate of the turbulent kinetic energy in the axisymmetric plane of the solar chimney for the different chimney diameters equal to d=0.1 m, d=0.13 m, d=0.16 m, and d=0.2 m. According to these results, it is found that the distribution of the dissipation rate of the turbulent kinetic energy is very weak in the axisymmetric plane except in the collector outlet. This fact is due to the important value of the viscous shear stress that appeared on the airflow in the solar chimney. Indeed, it is clear that the dissipation rate of the turbulent kinetic energy increases with the increase of the chimney diameter. The difference is illustrated at the maximum value of the dissipation rate of the turbulent kinetic energy, which is equal to $\varepsilon=231$ m^2.s^{-3} with the diameter d=0.1 m, to $\varepsilon=314$ m^2.s^{-3} with the diameter d=0.13 m, to $\varepsilon=368$ m^2.s^{-3} with the diameter d=0.16 m and to $\varepsilon=402$ m^2.s^{-3} with the diameter d=0.2 m.

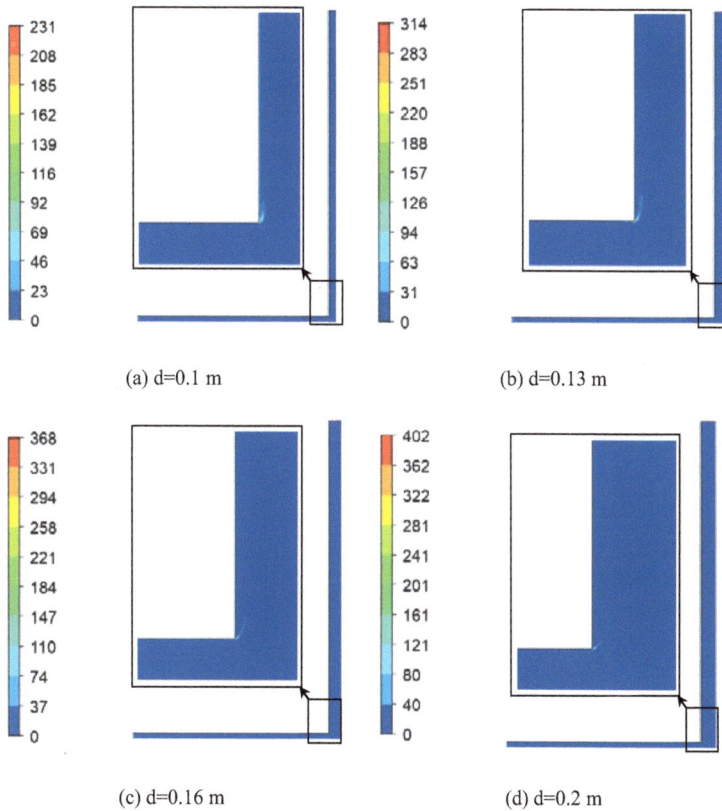

(a) d=0.1 m

(b) d=0.13 m

(c) d=0.16 m

(d) d=0.2 m

Fig. (4.56). Distribution of the dissipation rate of the turbulent kinetic energy for different chimney diameters.

4.4.6. Turbulent Viscosity

Fig. (**4.57**) shows the distribution of the turbulent viscosity in the axisymmetric plane of the solar chimney for the different chimney diameters equal to d= 0.1 m, d= 0.13 m, d= 0.16 m and d= 0.2 m. According to these results, the turbulent viscosity presents the same distribution for all diameters. It is low in the axisymmetric plane except in the chimney inlet since the magnitude velocity in this zone is maximum. The difference between the four cases is located in the chimney inlet when the highest value is found. Indeed, it has been observed that the turbulent viscosity increases with the increase of the chimney diameter. The maximum value is equal to μ_t =0.00292 kg.m^{-1}.s^{-1} with the diameter d= 0.1 m, to μ_t =0.00297 kg.m^{-1}.s^{-1} with the diameter d= 0.13 m, to μ_t =0.0026 kg.m^{-1}.s^{-1} with the diameter d= 0.16 and to μ_t =0.00273 kg.m^{-1}.s^{-1} with the diameter d= 0.2 m.

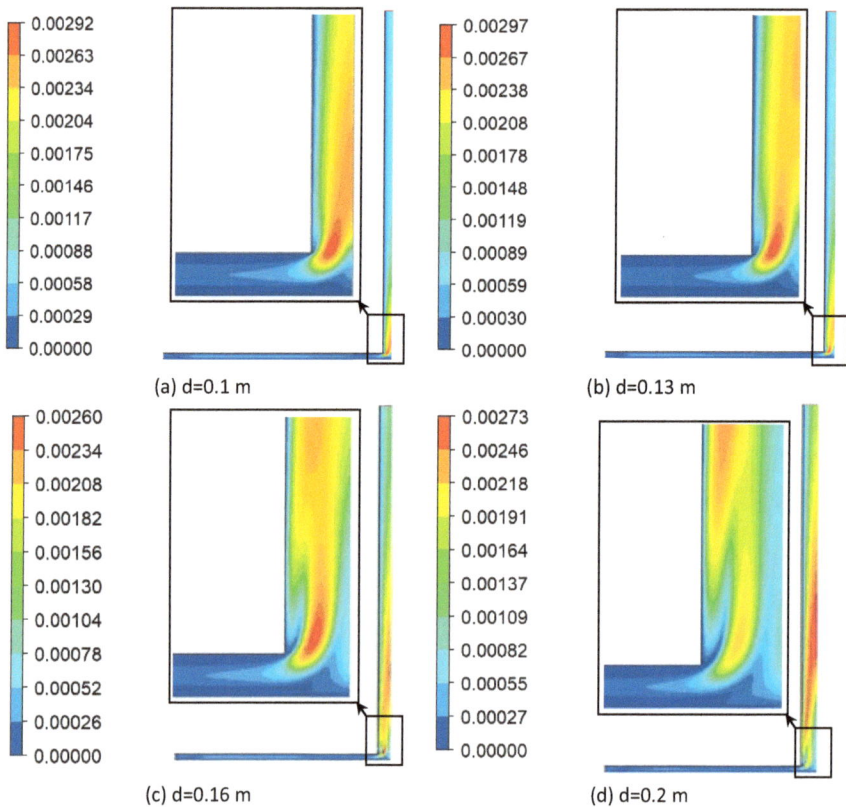

(a) d=0.1 m   (b) d=0.13 m

(c) d=0.16 m   (d) d=0.2 m

Fig. (4.57). Distribution of the turbulent viscosity for different chimney diameters.

4.5. Chimney Height Effect

In this section, we are interested in the study of the chimney height effect on the local results such as the magnitude velocity, the temperature, the static pressure, and the turbulence characteristic. The system considered is defined by a collector slope angle equal to $\theta=0°$, a chimney diameter $d=0.16$ m, a collector diameter $D=2.75$ m and a collector height $h=0.005$ m (Fig. **4.58**). Particularly, we have studied the solar chimney for four chimney height equal to H=1 m, H=2 m, H=3 m, and H=4 m. The different geometries are shown in Fig. (**4.59**).

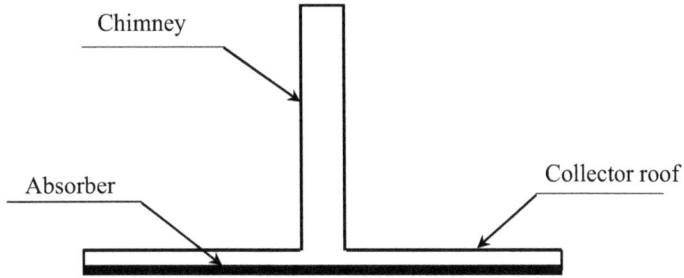

Fig. (4.58). Schematic of the system.

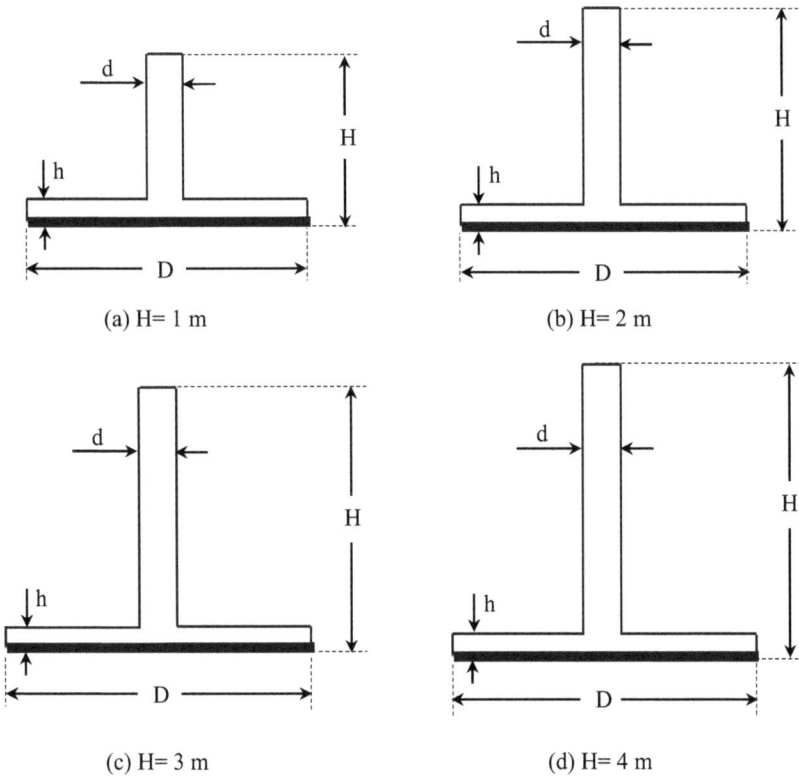

(a) H= 1 m

(b) H= 2 m

(c) H= 3 m

(d) H= 4 m

Fig. (4.59). Schematic of the different geometries.

4.5.1. Magnitude Velocity

Fig. (**4.60**) shows the variation of the maximum magnitude velocity for different chimney height equal to H=1 m, H=2 m, H=3 m, and H=4 m. According to these results, it is clear that the increase of the chimney height produces an increase in the maximum velocity. The elevation of the height produces an elevation of the differential pressure between the base and the outlet of the chimney. Figs. (**4.61**) and (**4.62**) show respectively the profiles of the magnitude velocity along the chimney axis and the profiles of the magnitude velocity in the collector. According to these results, it is clear that the velocity in the collector and the chimney axis increases with the increases of the chimney height. Indeed, the numerical results revealed that the chimney height is one of the most important parameters for the solar chimney design.

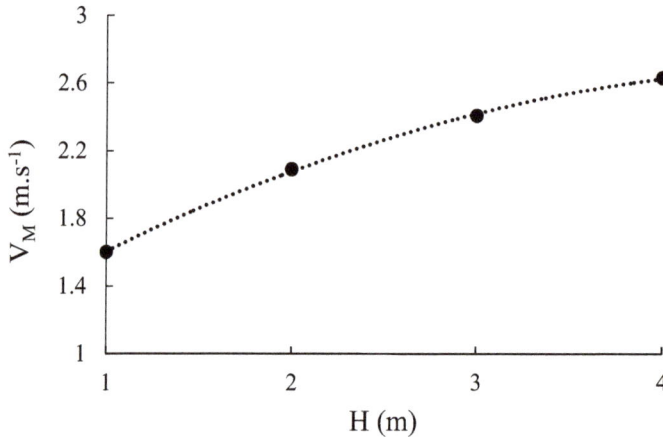

Fig. (4.60). Variation of the maximum magnitude velocity.

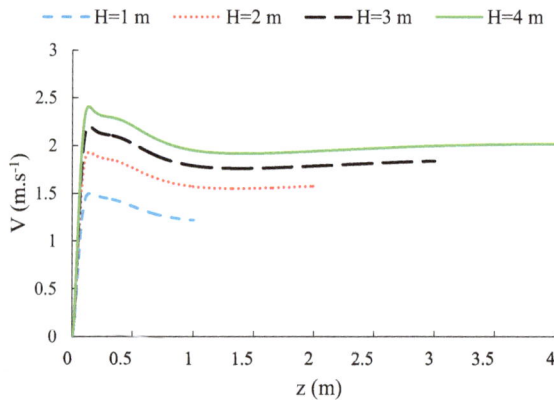

Fig. (4.61). Profiles of the magnitude velocity along the chimney axis.

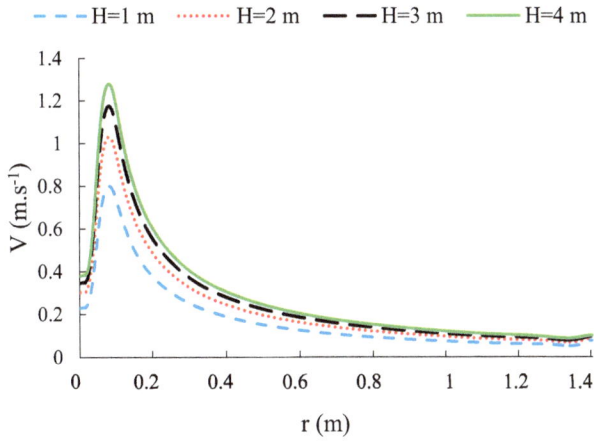

Fig. (4.62). Profiles of the magnitude velocity in the collector.

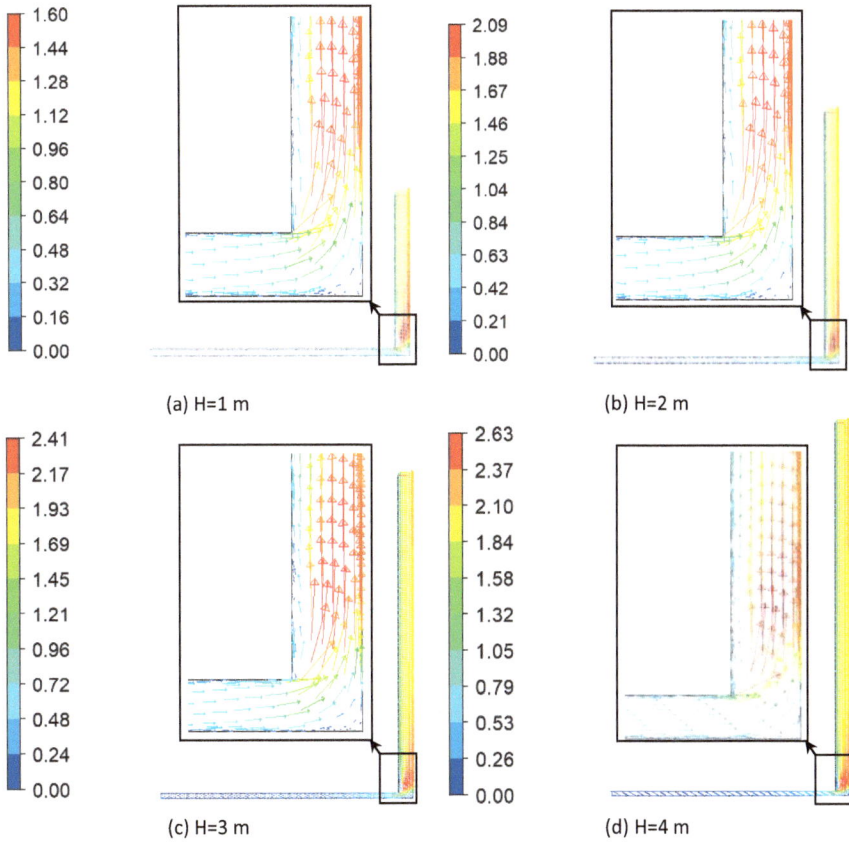

Fig. (4.63). Velocity fields for different chimney heights.

The distribution of the velocity fields and the magnitude velocity in the axisymmetric plane of the solar chimney for different chimney height equal to H=1 m, H=2 m, H=3 m, and H=4 m is shown respectively in Figs. (**4.63**) and (**4.64**). According to these results, it is clear that the magnitude velocity is low at the collector inlet and it increases gradually until the collector output. The acceleration zone is located at the chimney base for all heights. Besides, it can be observed that the magnitude velocity increases with the increase of the chimney height. The maximum value of the magnitude Velocity is equal to V=2.63 m.s^{-1} with H = 4 m, to V=2.41 m.s^{-1} with H =3 m, to V=2.09 m.s^{-1} with H =2 m and to V=1.6 m.s^{-1} with H=1 m.

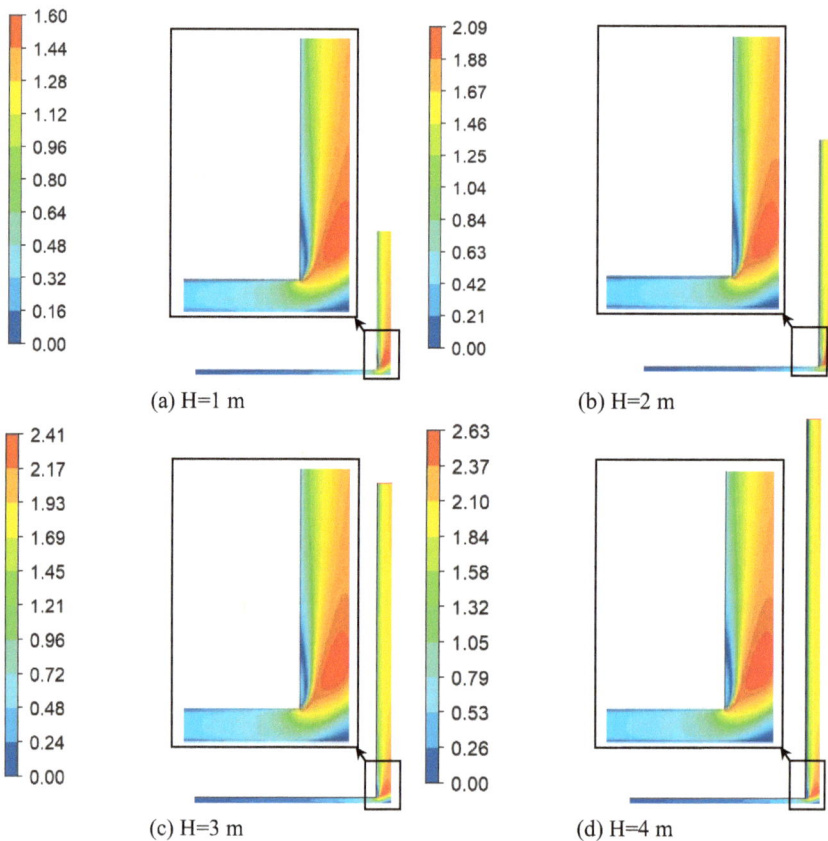

(a) H=1 m

(b) H=2 m

(c) H=3 m

(d) H=4 m

Fig. (4.64). Distribution of the magnitude velocity for different chimney heights.

4.5.2. Temperature

Fig. (**4.65**) displays the profiles of the temperature along with the collector. From these results, it is clear that the distribution of the temperature at the collector advanced contrariwise to the velocity. Indeed, if the chimney height upsurges, the volume of the system growths with constant solar radiation, then the temperature declines. Fig. (**4.66**) shows the distribution of the temperature in the defined plane of the solar chimney for different chimney height. From these results, it is clear that the variation in chimney height does not affect the distribution of temperature. Certainly, the maximum temperature value in the collector centerline drops with the rise of the tower height. But, for each height value, the airflow in its advancement through the collector continues to absorb heat making, thus, its temperature increases.

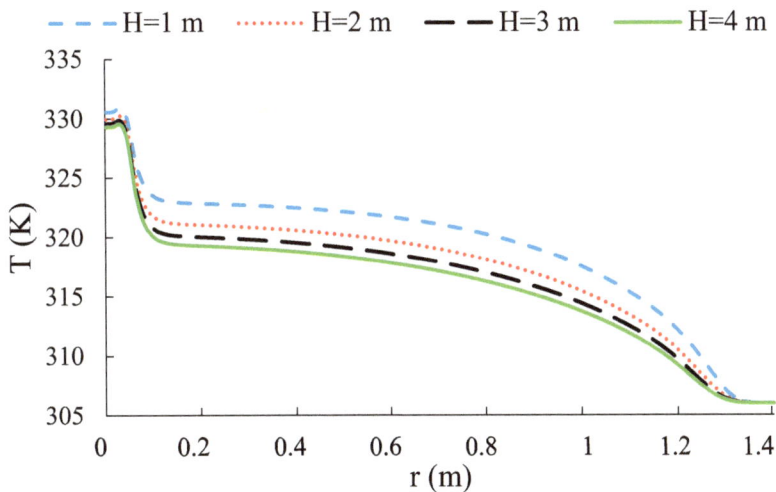

Fig. (**4.65**). Profiles of the temperature in the collector.

4.5.3. Static Pressure

Fig. (**4.67**) displays the static pressure profiles along the chimney axis for different chimney height equal to H=1 m, H=2 m, H=3 m, and H=4 m. From these results, it is clear that the static pressure presents a similar distribution for all chimney height. The depression value is extreme at the chimney entrance where the velocity is maximum. Formerly, the static pressure rises and ranges the atmospheric pressure value at the chimney outlet. Then, this pressure change is liable for the airflow along with the chimney. Besides, it has been observed that the increase of the chimney height leads to increase the depression at the base of the chimney. This point is proved from the static pressure distribution in the defined plane of the chimney axis for different chimney heights as exposed in Fig.

(4.68). According to these results, it is clear that the depression zone is located in the base of the chimney. Elsewhere, the static pressure increases until reaching the atmospheric pressure in the inlet of the collector and the outlet of the chimney. In these conditions, the difference is clear at the maximum value of the static pressure is equal to p=2.05 Pa with H=1 m, to p=3.42 Pa with H=2 m, to p=4.47 Pa with H=3 m and to p=5.30 Pa with H=4 m.

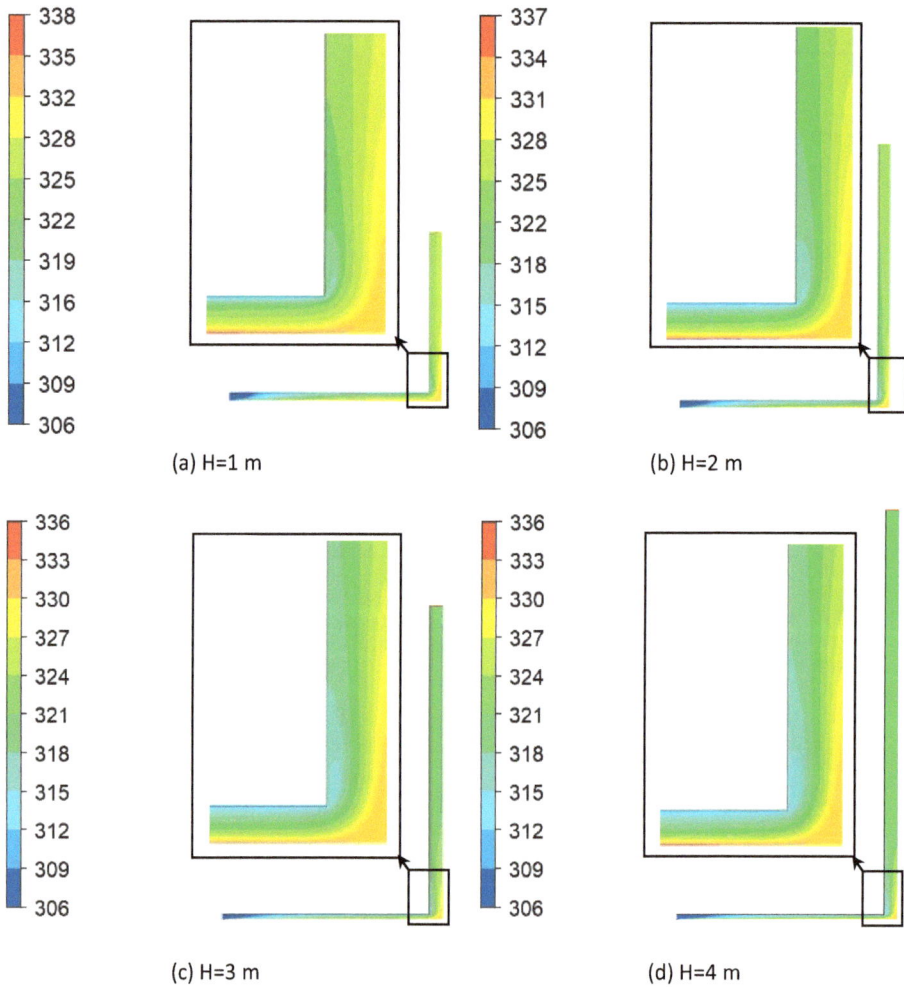

(a) H=1 m   (b) H=2 m

(c) H=3 m   (d) H=4 m

Fig. (4.66). Distribution of the temperature for different chimney heights.

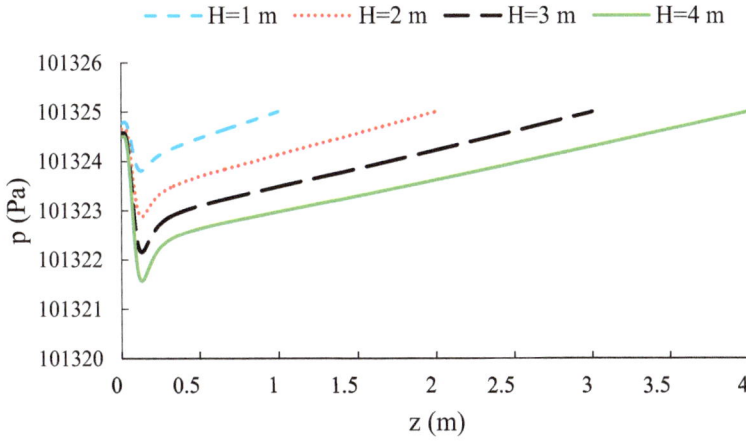

Fig. (4.67). Profiles of the static pressure along the chimney axis.

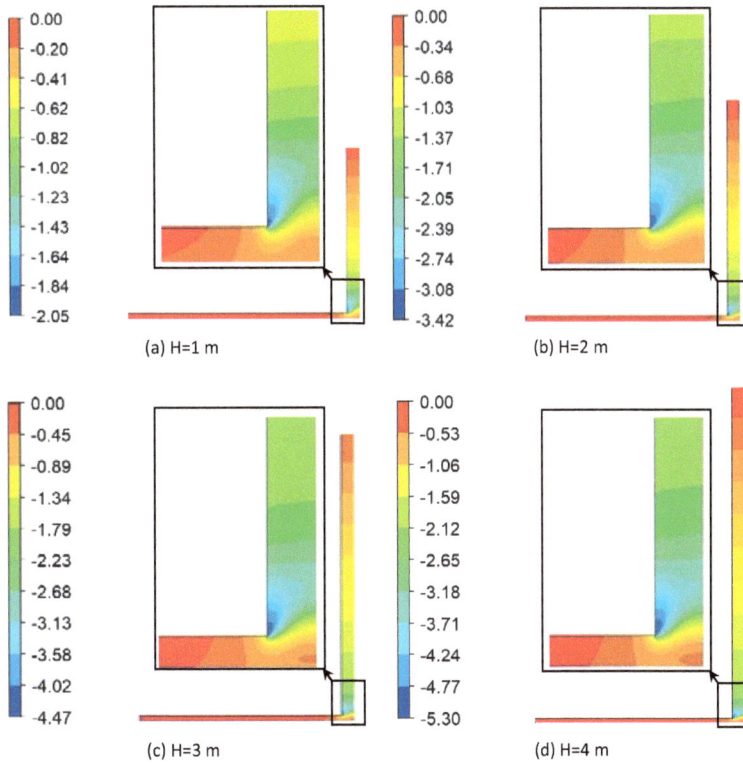

(a) H=1 m   (b) H=2 m

(c) H=3 m   (d) H=4 m

Fig. (4.68). Distribution of the static pressure for different chimney heights.

4.5.4. Turbulent Kinetic Energy

Fig. (**4.69**) shows the distribution of the turbulent kinetic energy in the axisymmetric plane of the solar chimney for the different chimney heights equal to H=1 m, H=2 m, H=3 m, and H=4 m. According to these results, the turbulent kinetic energy presents the same distribution for all diameters. The turbulent kinetic energy is low in the axisymmetric plane except in the chimney inlet. There may be a small difference between the distributions of the turbulent kinetic energy for each height. This difference is located in the chimney inlet when the highest value is found in the fourth case with the height H=4 m. The maximum value of the turbulent kinetic energy is equal to k=0.173 $m^2.s^{-2}$ with the height H=1 m, to k=0.279 $m^2.s^{-2}$ with the height H=2 m, to k=0.367 $m^2.s^{-2}$ with the height H=3 m and to k=0.438 $m^2.s^{-2}$ with the height H=4 m.

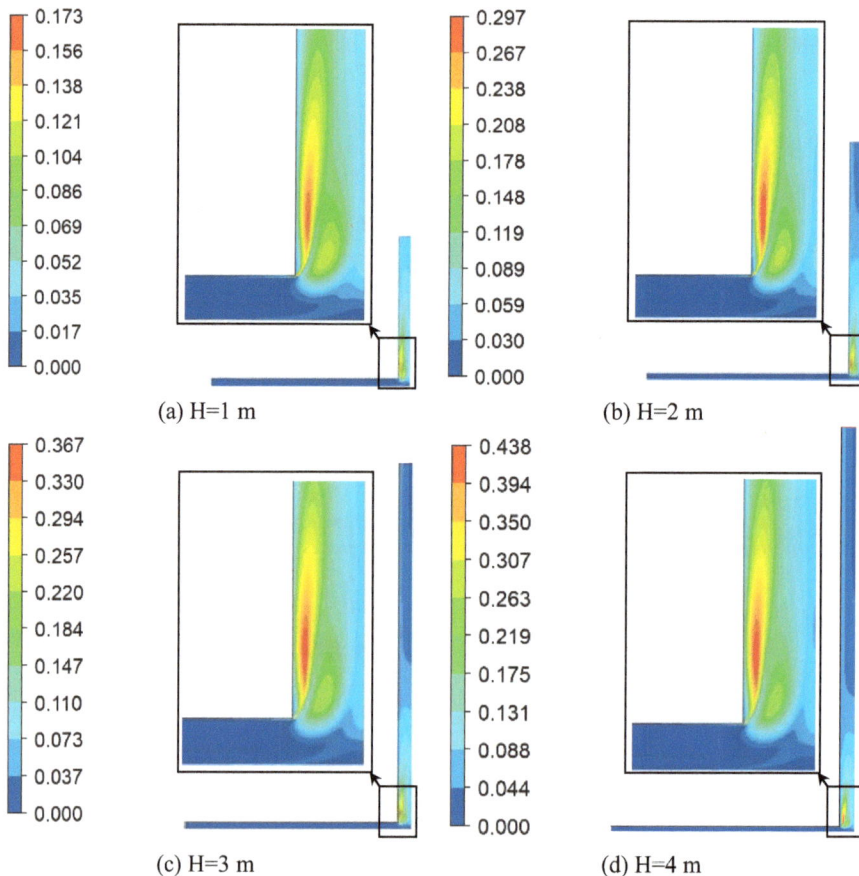

(a) H=1 m

(b) H=2 m

(c) H=3 m

(d) H=4 m

Fig. (4.69). Distribution of the turbulent kinetic energy for different chimney heights.

4.5.5. Dissipation Rate of the Turbulent Kinetic Energy

Fig. (**4.70**) shows the distribution of the dissipation rate of the turbulent kinetic energy in the axisymmetric plane of the solar chimney for the different chimney heights equal to H=1 m, H= 2 m, H=3 m, and H=4 m. The dissipation phenomena are due to the turbulent kinetic energy when it was converted to internal energy for example the thermal energy. Rendering to these results, the difference of the chimney height does not affect the distribution of the dissipation rate of the turbulent kinetic energy, which is found to be very weak in the axisymmetric plane excluding in the collector exit. Certainly, it is clear that the dissipation rate of the turbulent kinetic energy upsurges with the increase of the chimney height.

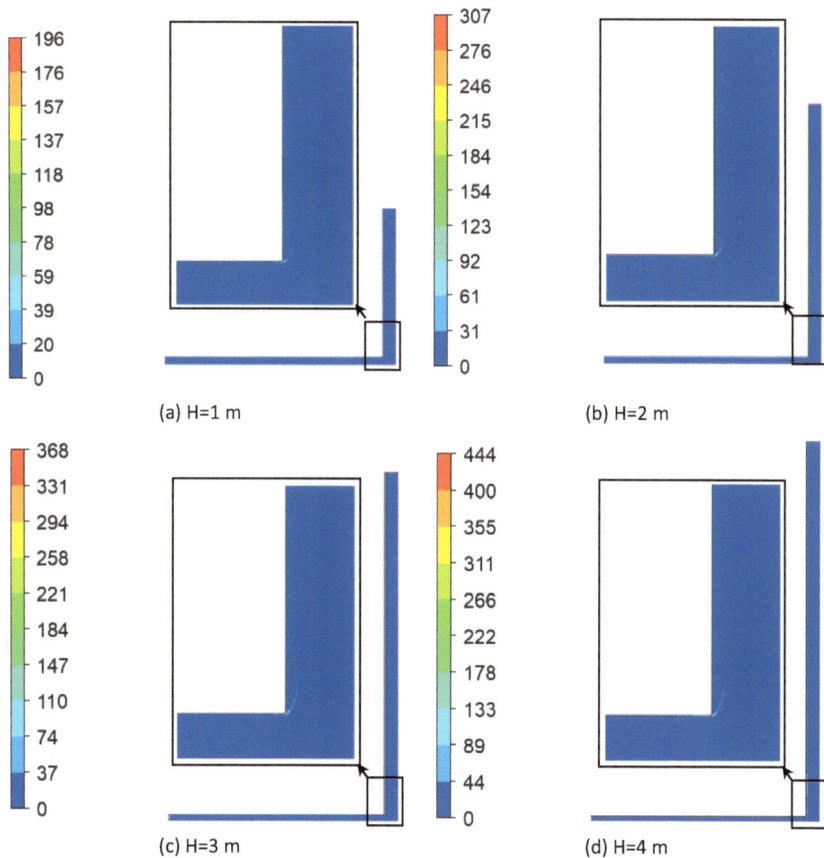

Fig. (4.70). Distribution of the dissipation rate of the turbulent kinetic energy for different chimney heights.

4.5.6. Turbulent Viscosity

Fig. (**4.71**) displays the distribution of the turbulent viscosity in the defined plane of the solar chimney for the diverse chimney heights equal to H=1 m, H=2 m, H=3 m, and H=4 m. According to these results, the turbulent viscosity presents the same distribution for all cases. In these conditions, it is clear that the turbulent viscosity is low in the axisymmetric plane except in the chimney inlet. Indeed, it may appear a small difference between the distributions of the turbulent viscosity for each height. This change is set in the chimney inlet when the maximum value is found in the fourth case with height H=4 m. The maximum value of the turbulent viscosity is equal to μ_t =0.00192 kg.m^{-1}.s^{-1} with the height H=1 m, to μ_t =0.00134 kg.m^{-1}.s^{-1} with the height H=2 m, to μ_t =0.0026 kg.m^{-1}.s^{-1} with the height H=3 m and to μ_t =0.00278 kg.m^{-1}.s^{-1} with the height H=4 m.

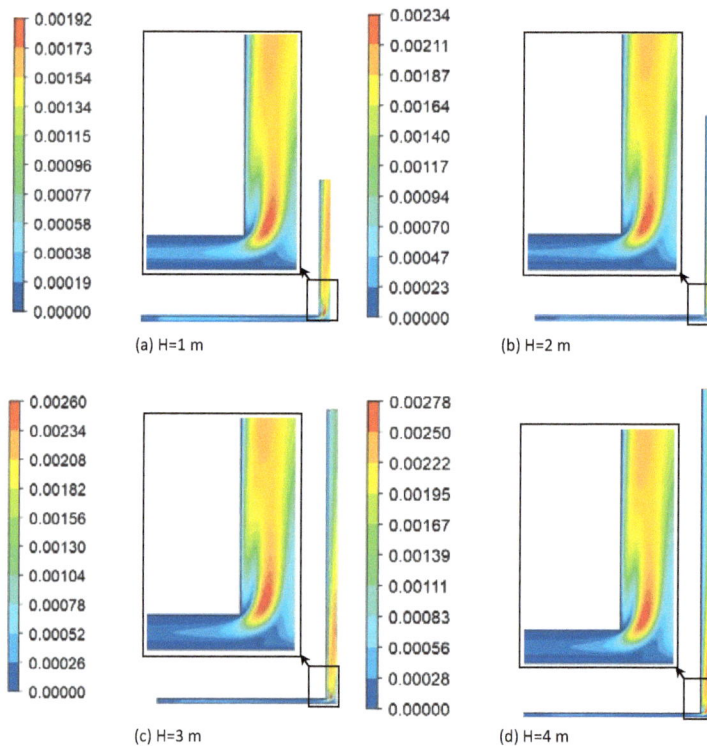

Fig. (4.71). Distribution of the turbulent viscosity for different chimney heights.

4.6. Chimney Forms Effect

In this part, we studied the effect of the chimney form on the system performance. The system considered is defined by a collector slope angle equal to θ=0°, a collector diameter D=2.75 m, and a collector height h=0.005 m (Fig. **4.72**). Particularly, we have studied the solar chimney for four chimney forms consisting of an opposing cone, a cone, an inverted cone, and a tube. The different geometries are shown in Fig. (**4.73**).

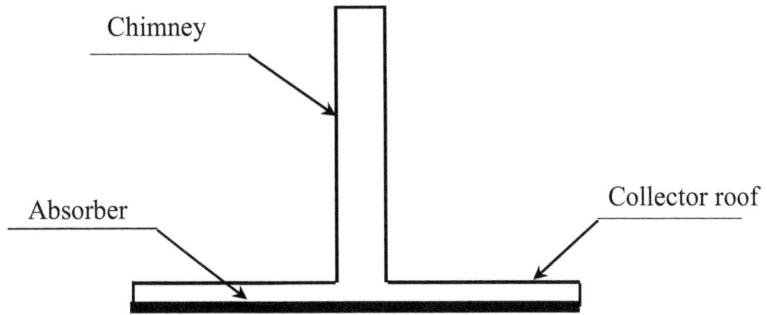

Fig. (4.72). Schematic of the system.

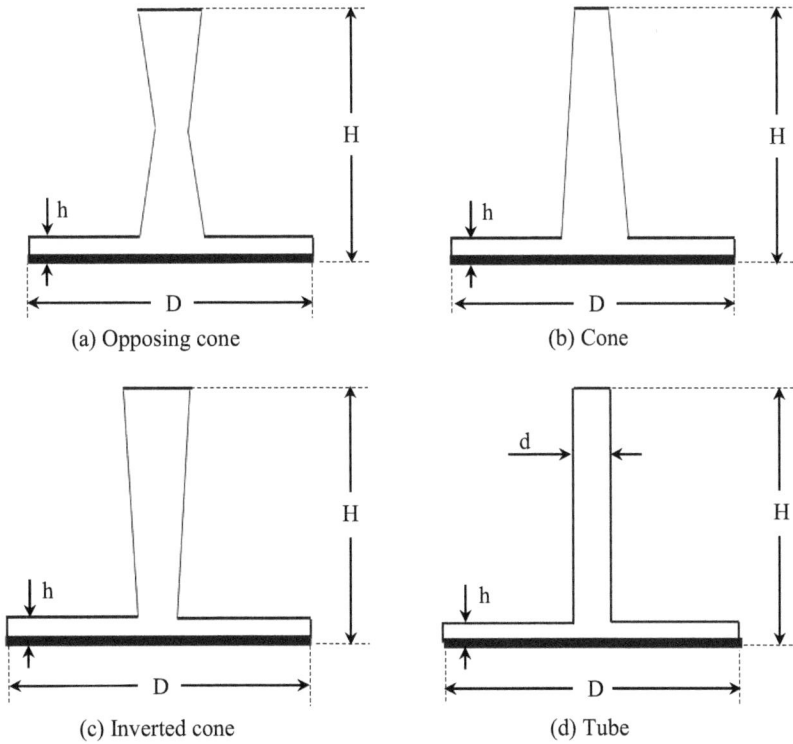

Fig. (4.73). Schematic of the different geometries.

4.6.1. Magnitude Velocity

Table **4.4** illustrates the maximum magnitude velocity value for different chimney forms. According to these results, it is clear that the chimney forms directly affect the maximum velocity value and system performance. Indeed, it is illustrated that the installation by the form "inverted cone" occupies the best performance because the maximum air velocity reaches 3.5 m.s^{-1} which is higher than the other forms. Besides, the distribution of the velocity fields, the magnitude velocity in the axisymmetric plane of the solar chimney for different chimney forms is shown respectively in Figs. (**4.74** & **4.75**). According to these results, it is clear that the velocity evolved inversely to the static pressure. To understand this phenomenon, we can refer to Bernoulli's theorem, where fluid velocity must increase as it passes through a constriction following the principle of mass continuity. While its static pressure must decrease under the principle of conservation of mechanical energy. Thus, a drop in pressure balances gains in kinetic energy a fluid may accrue due to its increased velocity through a constriction. By measuring the change in pressure, the flow rate can be determined, as in various flow measurement devices.

Table 4.4. Maximum velocity.

Forms	Tube	Cone	Inverted Cone	Opposing Cone
Maximum Velocity (m.s^{-1})	2.405	2.338	3.513	2.7168

4.6.2. Static Pressure

Fig. (**4.76**) shows the distribution of the static pressure in the axisymmetric plane of the solar chimney for different chimney forms on an opposing cone, a cone, an inverted cone, and a tube. According to these results, it is clear that the depression value is maximum at the chimney inlet where the velocity is maximum. It has been observed that the static pressure is maximum for the inverted cone form since the magnitude velocity is maximum. Besides, it is illustrating that the depression zone is located in the base of the chimney. Elsewhere, the static pressure increases until reaching the atmospheric pressure in the inlet of the collector. In these conditions, the difference is clear at the maximum depression value of the static pressure is equal to p=4.94 Pa with the opposing cone form, to p=2.39 Pa with the cone form, to p=8.87 Pa with the inverted cone form and to p=4.47 Pa with the tube form.

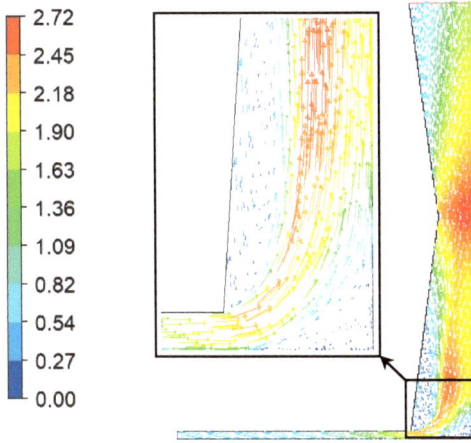

(a) Opposing cone   (b) Cone

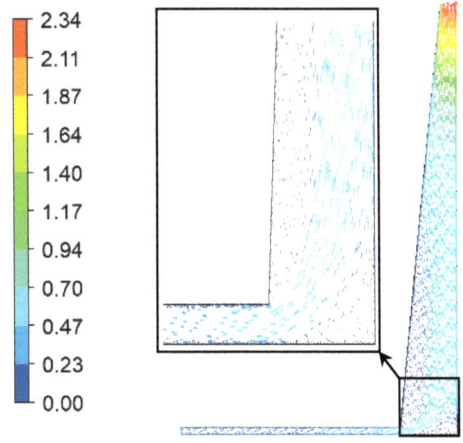

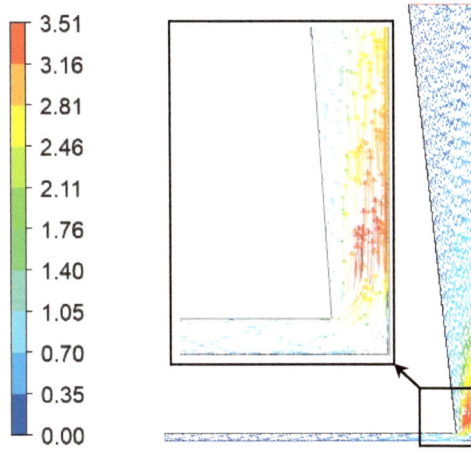

(c) Inverted cone   (d) Tube

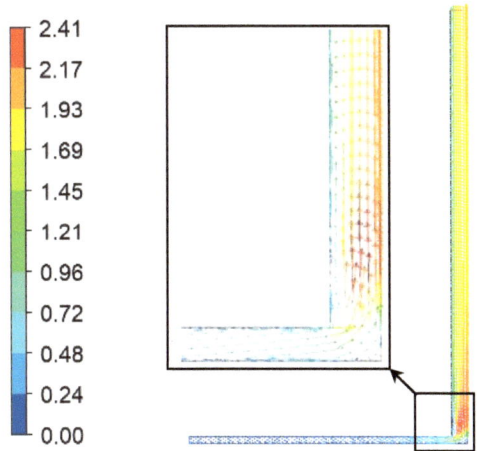

Fig. (4.74). Velocity fields for different chimney forms.

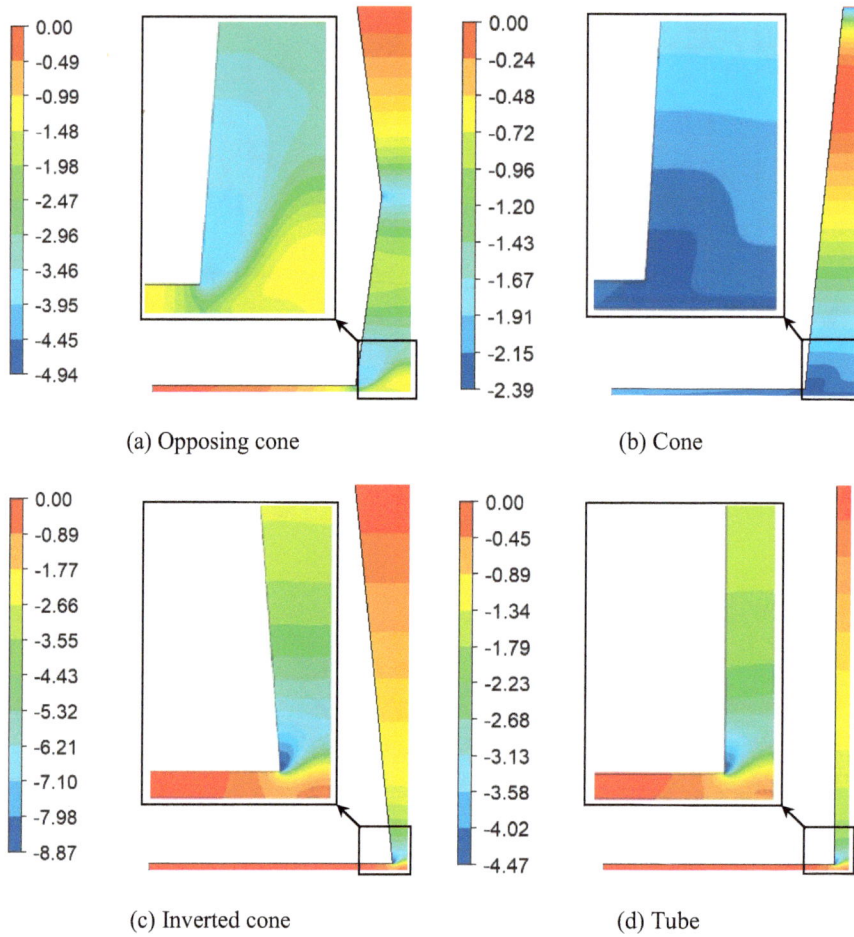

(a) Opposing cone

(b) Cone

(c) Inverted cone

(d) Tube

Fig. (4.75). Distribution of the magnitude velocity for different chimney forms.

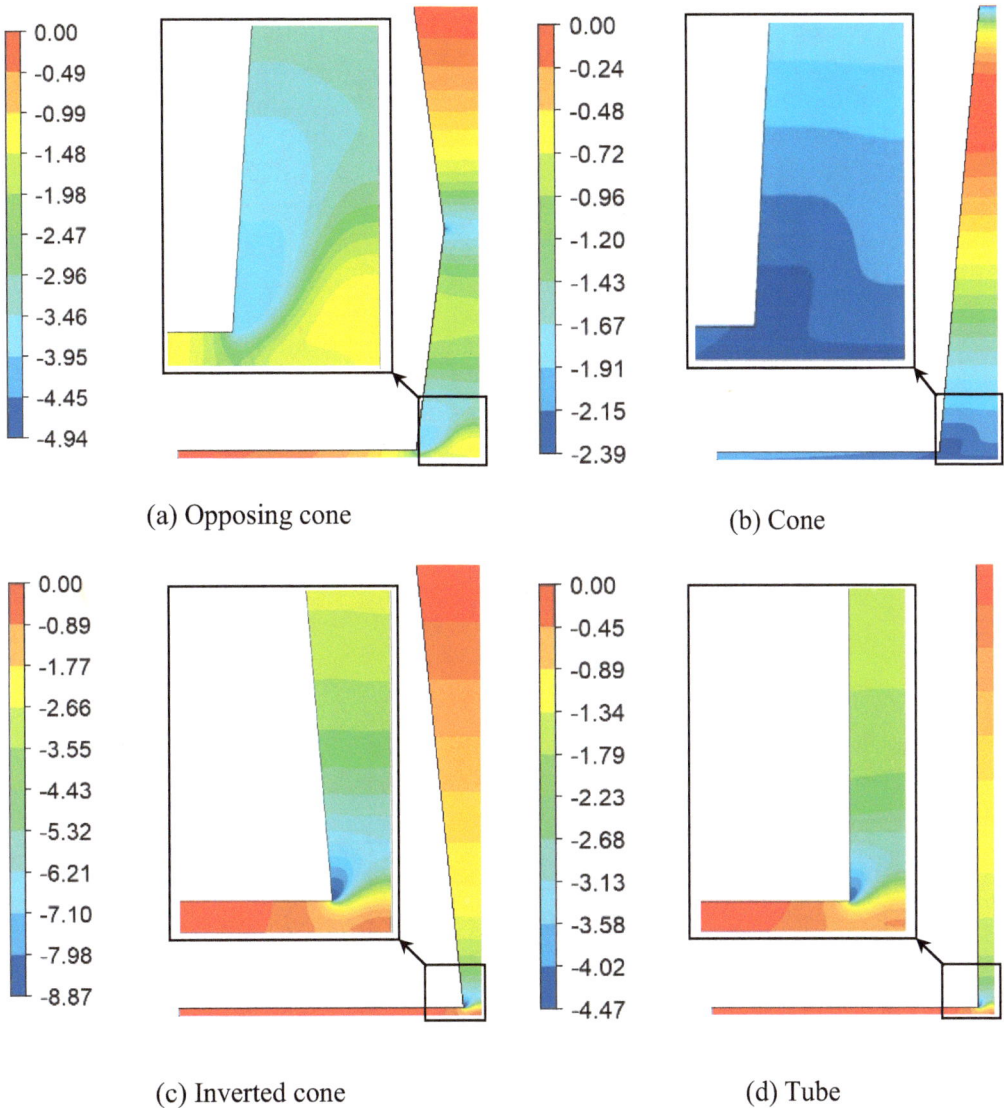

(a) Opposing cone   (b) Cone

(c) Inverted cone   (d) Tube

Fig. (4.76). Distribution of the static pressure for different chimney forms.

4.6.3. Turbulent Kinetic Energy

Fig. (**4.77**) shows the distribution of the turbulent kinetic energy in the axisymmetric plane of the solar chimney for the different chimney forms consisting of an opposing cone, a cone, an inverted cone, and a tube. According to these results, the turbulent kinetic energy presents the same distribution for all heights. It is low in the axisymmetric plane except in the chimney inlet since the magnitude velocity in this zone is maximum. The difference between the four cases is located in the chimney inlet when the highest value is found. It has been observed that the turbulent kinetic energy is maximum for the inverted cone form since the magnitude velocity is maximum. The maximum value of the turbulent kinetic energy is equal to k=0.334 m².s⁻² with the opposing cone form, to k=0.123 m².s⁻² with the cone form, to k=0.843 m².s⁻² with the inverted cone form and to k=0.367 m².s⁻² with the tube form.

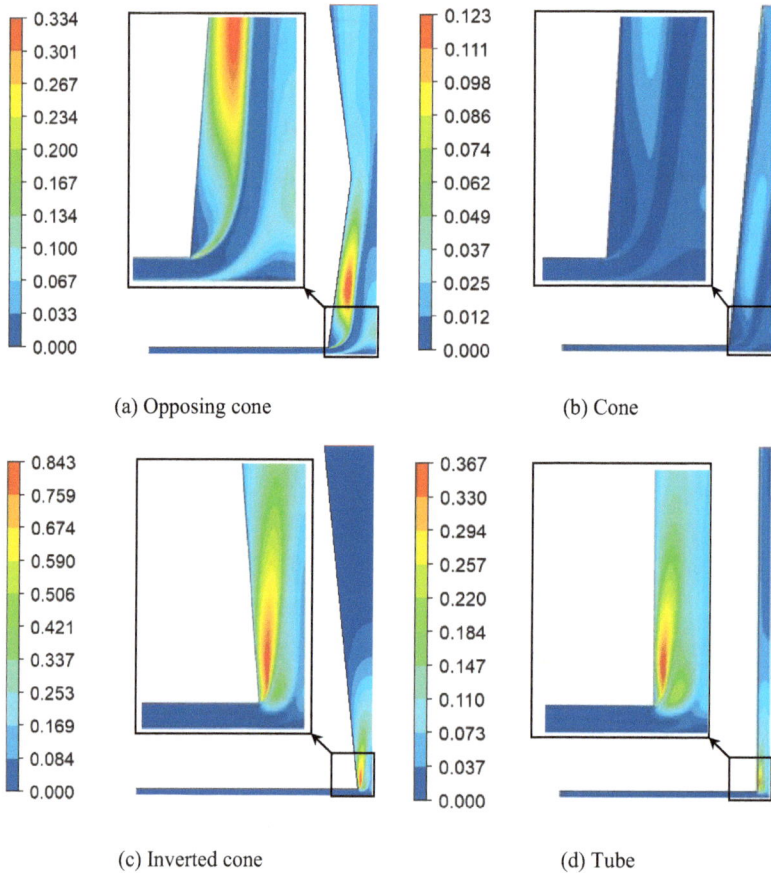

(a) Opposing cone

(b) Cone

(c) Inverted cone

(d) Tube

Fig. (4.77). Distribution of the turbulent kinetic energy for different chimney forms.

4.6.4. Dissipation Rate of the Turbulent Kinetic Energy

Fig. (**4.78**) shows the distribution of the dissipation rate of the turbulent kinetic energy in the axisymmetric plane of the solar chimney for the different chimney forms consisting of an opposing cone, a cone, an inverted cone, and a tube. According to these results, it is found that the distribution of the dissipation rate of the turbulent kinetic energy is very weak in the axisymmetric plane except in the collector outlet. This fact is due to the important value of the viscous shear stress that appeared on the airflow in the solar chimney. Indeed, it is clear that the dissipation rate of the turbulent kinetic energy is maximum for the inverted cone form since the magnitude velocity is maximum. The difference is illustrated at the maximum value of the dissipation rate of the turbulent kinetic energy, which is equal to $\varepsilon=92$ m^2.s^{-3} with the opposing cone form, to $\varepsilon=130$ m^2.s^{-3} with the cone form, to $\varepsilon=596$ m^2.s^{-3} with the inverted cone form and to $\varepsilon=368$ m^2.s^{-3} with the tube form.

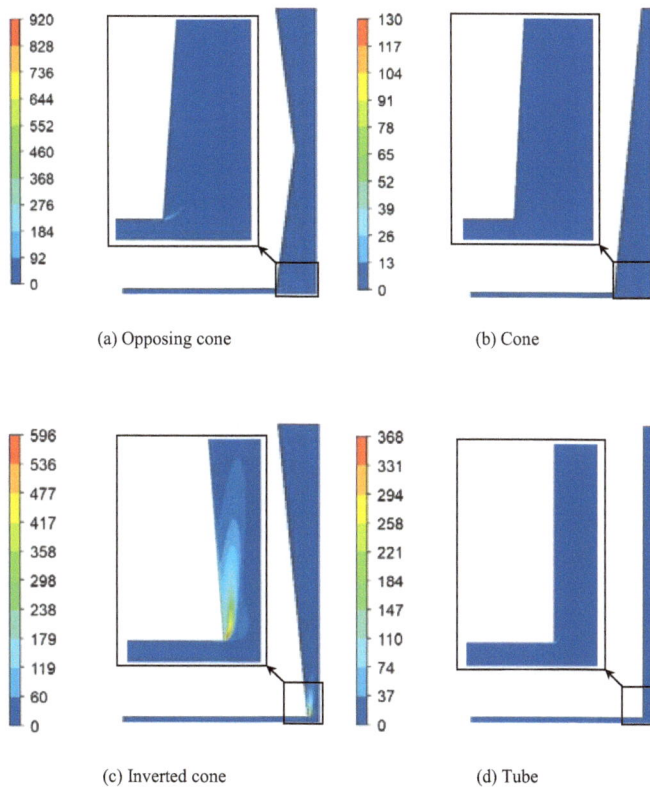

(a) Opposing cone   (b) Cone

(c) Inverted cone   (d) Tube

Fig. (4.78). Distribution of the dissipation rate of the turbulent kinetic energy for different chimney forms.

4.6.5. Turbulent Viscosity

Fig. (**4.79**) displays the distribution of the turbulent viscosity in the defined plane of the solar chimney for the different chimney shapes consisting of an opposing cone, a cone, an inverted cone, and a tube. From these results, the turbulent viscosity gives a different distribution for each form. The difference between the four cases is clear in the chimney since the velocity and the static pressure are not the same distribution. It has been observed that the turbulent viscosity is maximum for the inverted cone form since the magnitude velocity is maximum in this form. In these condition, the maximum value is equal to μ_t =0.00599 kg.m^{-1}.s^{-1} with the opposing cone form, to μ_t =0.00316 kg.m^{-1}.s^{-1} with the cone form, to μ_t =0.00748 kg.m^{-1}.s^{-1} with the inverted cone form and to μ_t =0.0026 kg.m^{-1}.s^{-1} with the tube form.

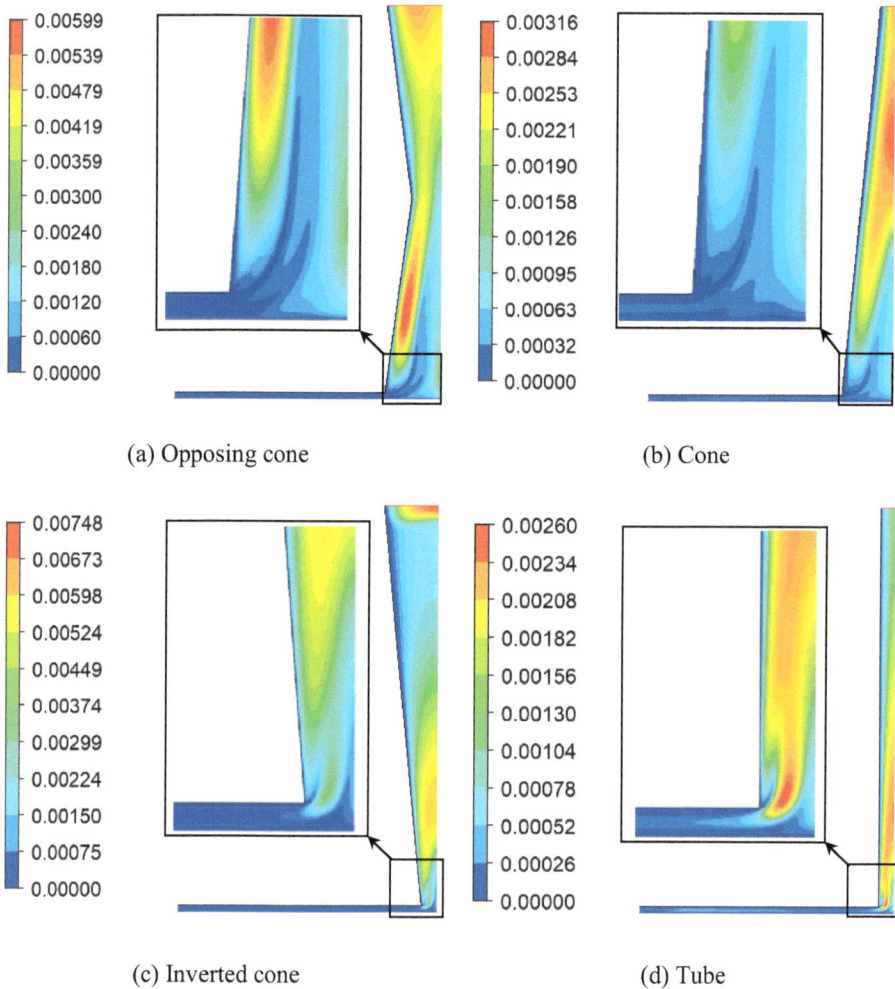

(a) Opposing cone (b) Cone

(c) Inverted cone (d) Tube

Fig. (4.79). Distribution of the turbulent viscosity for different chimney forms.

5. OPTIMUM GEOMETRY CHOICE

We have encountered many constructive problems during the realization. The numerical results show that the maximum velocity obtained is for the collector height equal to h=0.05 m, the chimney diameter equal to d=0.2 m, the collector slope angle equal to θ=0°, the collector diameter equal to D=4 m and the chimney height equal to H=4 m. The chimney height cannot withstand the huge force of the wind. Therefore, we choose the slightly less height geometry (H=3 m). Besides, there is a difficulty in terms of chimney diameter. Indeed, the maximum air velocity is obtained for a diameter d=0.2 m, but this diameter is not available. Therefore, we choose the diameter d = 0.16 m. Taking into account the difficulties, the results of numerical simulations have shown that the collector height equal to h=0.05 m, the collector diameter is D=2.75 m, the chimney diameter is d=0.16 m and the chimney height is H=3 m are the optimal dimensions for the design of the solar chimney power plant. The 3D model for our prototype is shown in Fig. (**4.80**).

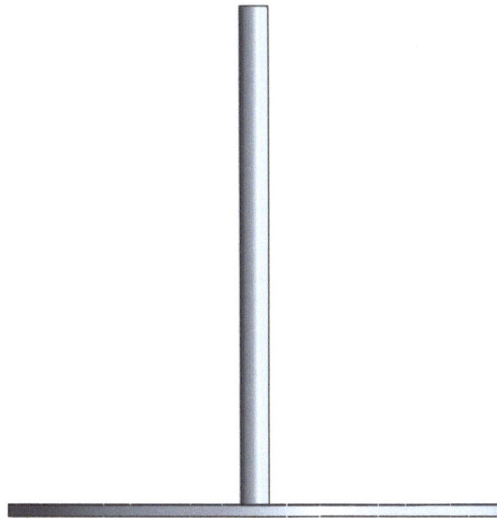

Fig. (4.80). 3D model of our prototype.

6. VALIDATION WITH EXPERIMENTAL RESULTS

The experimental facts considered for the validation of the numerical model were reported on May 28, 2016. Also, two important parameters are engaged to achieve validation. The first being the temperature distribution inside the collector at the height of 1 cm, which is obtained through 5 positions with different sensors with a 30 cm distance between each of them. Figs. (**4.81**) and (**4.82**) present, respectively the profiles of the temperature and the magnitude velocity along the collector.

These differences between the experimental and numerical results could have several reasons such as measurement error, changing the direction of radiation, which can cause the shadow of the sensors and the environmental condition like the wind. Also, the difference between the numerical and experimental results could be for some assumption and simplification in the numerical model like assume constant of the heat transfer coefficient.

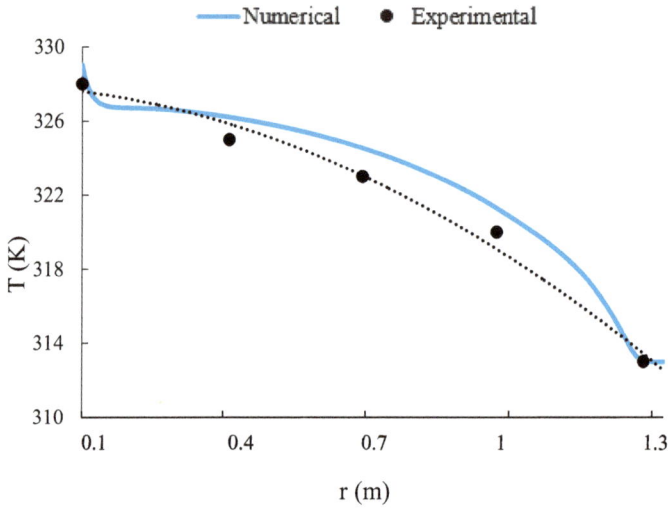

Fig. (4.81). Temperature distribution along the collector radius on May 28, 2016 at 13:00.

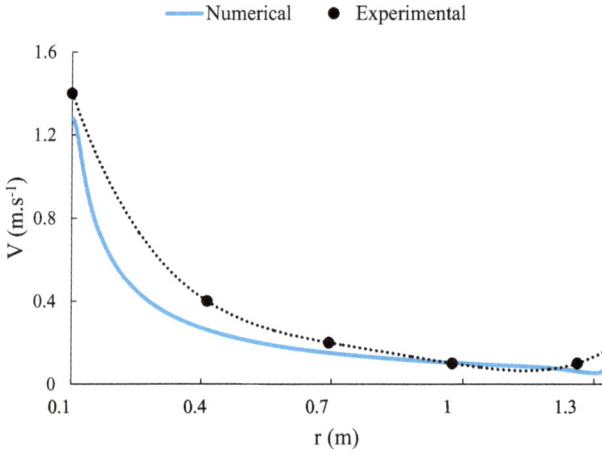

Fig. (4.82). Velocity distribution along the collector radius on May 28, 2016 at 13:00.

CONCLUSION

In this chapter, the effect of different geometrical parameters such as the collector height, the collector diameter, the chimney diameter, the chimney height, and the collector slope angle were studied. The numerical profiles for the temperature, the magnitude velocity, the static pressure, and the turbulent characteristics in the solar chimney power planet are shown for different geometrical parameters. In these conditions, it has been noted that the sizes of the tower affect the airflow, either by an increase or decrease in mass flow. Besides, we have managed to perform a numerical model approaching reality, and compare it with our experimental results to validate it. Therefore, we fixed the various parameters for use in the next chapter.

CHAPTER 5

Experimental Study

1. INTRODUCTION

There are many configuration sizes of the small-scale solar chimney in the world, because of various meteorological and geographic conditions. The main objective of their works is to test the heat transfer performance of the collector and the power output. However, our work is based on the numerical study and numerical optimization of geometric parameters to testing the performance of the prototype with meteorological data. In the first part, we are interested in the experimental characterization of a solar chimney with specific instrumentation. In particular, we present a detailed description of the realization and the various manipulations made to the study. A discussion of results is then presented at the end of this chapter.

2. PROTOTYPE PRESENTATION

The prototype consists of a solar chimney of a small scale that we have conducted to highlight the physical phenomena studied on the numerical system, including the natural convection phenomena. After we respected the dimensions that we choose in the precedent chapter, we have created a design that is simple, effective, and rigid to construct our prototype of the solar chimney by using simple and available tools. Fig. (**5.1**) shows a schematic view of the experimental solar chimney for power generation situated on the experimental site. It consists of five major components, such as:

· Collector (matrix with roof)

· Chimney

· Support of chimney

· Turbine generator

· Absorber

Haythem Nasraoui, Moubarek Bsisa & Zied Driss

The dimensions of our prototype are represented in Fig. (**5.2**), consisting of a collector with a diameter equal to D=2.75 m and a height equal to h=5 cm. The chimney has H=3 m in tall and d=16 cm in diameter. The slope angle of the collector is equal to q=0°. Besides, the stage of the realization of our prototype is given by appendix 2.

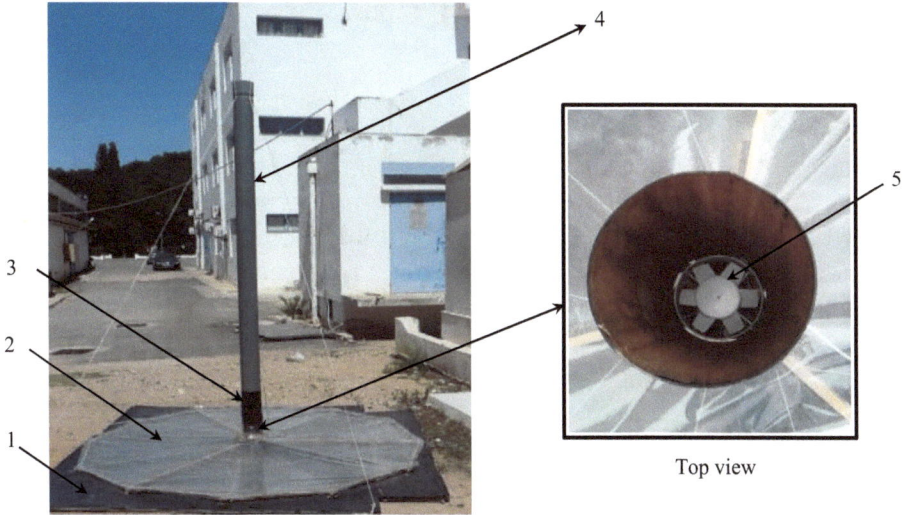

Fig. (5.1). Our prototype.

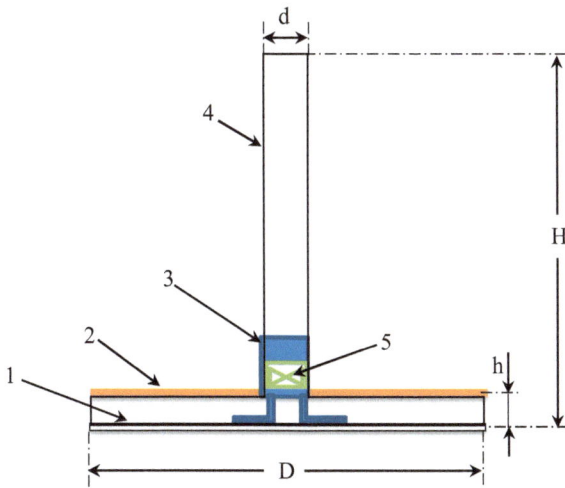

Fig. (5.2). Schematic of our prototype.

Rep.	Designation
1.	Absorber
2.	Collector Roof

(Table 5.2) cont.....

Rep.	Designation
3.	Chimney Support
4.	Chimney
5.	Turbine Generator

2.1. Collector

The collector must ensure good insulation of the front face of the sensor exposed to the solar radiation. For that, it must let pass the maximum of the energy of the exterior towards the absorber while being opposed to heat exchange reverse whatever their nature. In the same way, it must be transparent with the solar radiation but opaque with the infrared radiation. The second significant parameter by carrying out this chimney is the distance between the collector and the shock absorber. The larger this distance, the more the convective exchange is significant. For this, we must reduce the thickness as much as possible to reduce to the minimum the loss of heat outside. The most usual material is glass considering the factor of transmissivity for radiation solar is 0.85 without using it as cover, considering which it is very fragile. We replace glass by other materials that transmit less radiations but is easy with cutting and deformation as an adhesive film (tarpaulin) shown in Fig. (5.3). The thickness of the tarpaulin is another important parameter because if the thickness is significant than the transmissivity is low. For this reason, we have used a polyethylene with low density.

2.2. Chimney and Support

We take a PVC pipe for the chimney. It has H=3 m in tall, d=16 cm in internal diameter and 3 mm in thickness. We choose PVC because of its strength, maneuverability on change, good thermal resistivity, lightweight, resistivity to climatic factors (rain, wind, sun), and availability. The pipe is fixed on a support. This support should be a cylindrical boring with 16.5 cm in internal diameter, where it supported the PVC pipe of 16.3 cm in external diameter. Because we do not have a cylindrical shape with 16.5 cm in diameter in our workshop, we will use and weld a steel dish to obtain a cylindrical shape. To fix this cylindrical boring on the wooden platform, we weld three square on the external surface of the cylinder. We choose three square only because of decrease friction with air and increase the area in which the air enter the PVC pipe. The distance between the wooden platform and the inlet cylinder is 15.3 cm because we will have moved the pipe to change the slope of the collector. Fig. (5.4) shows the system chimney and support.

Fig. (5.3). Collector of our prototype.

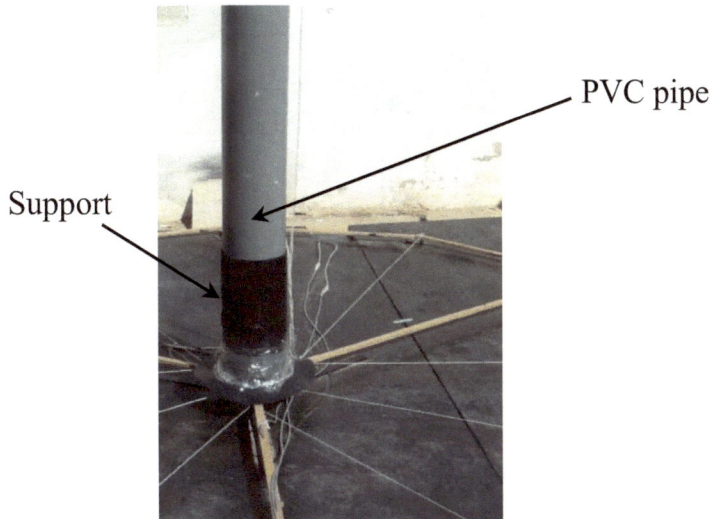

Fig. (5.4). Chimney of our prototype.

2.3. Turbine Generator

The role of the turbine is to transform the air energy into mechanical energy to rotate an alternator. The generator, in turn, transforms the mechanical energy into electrical energy. We choose a turbine type available and consist of the dimensions of our prototype, which is presented in Fig. (**5.5**). According to the numerical results, the maximum velocity appears at the chimney inlet. So, we placed the turbine in the chimney base, as shown in Fig. (**5.2**).

Fig. (5.5). Turbine generator.

2.4. Absorber

The absorber is a combination of a wooden platform with an 8 mm thickness with a black layer of paint. The interest of the black pigment is that it absorbs a good part of the luminous spectrum and reflects very little of it. The wood was selected because it is a good thermal insulator and causes to enhance the performance of the system. The platform on which it is fixed, the metallic support consists of three wooden tables for strength and it is painted in black to promote the "absorption of sunlight.

3. INSTRUMENTATION

To study the behavior of our system, we measured the temperature, the velocity, the current, and the voltage. Therefore, we will present the available instruments witch we have used for the measurement.

3.1. Velocity Measuring

To measure the air velocity in our system, it was necessary to have a measuring instrument with the best precision possible. Most sensors used for the measurement of air velocity are generally hot-wired since these are simple and offer the greatest flexibility of use and high resolution for use in different applications. Among them, the choice was now directed to a hot-wire anemometer (Fig. **5.6**) to make measurements in the tenth meter per second. The principle of these devices is to heat, by Joule effect, an element (wire or film) whose resistance depends on temperature. This element, placed in a fluid flow, convection cooling; its temperature, and thus its strength, then is primarily related to the fluid velocity, but also its temperature and physical characteristics of the fluid regulating the heat transfer between the element and the surrounding environment. An anemometer with an accuracy of ±0.01 m.s-1 is used to measure the velocity of air flow.

3.2. Temperature Measuring

Very often in electronics, we would need to measure temperature, not just in one spot, but simultaneously in two or more spots. Besides, log those measurements, so that we can plot temperature change over time. Moreover, differential temperature measurements could be extremely useful in some cases, especially for the thermal design of electronic projects. The Uni-Trend Model UT325 Thermometer is microprocessor-based, digital thermometers designed to use external J, K, T, E, R, S, and N-type thermocouples as temperature sensors (Fig. **5.7**). This apparatus has a Software and USB data cable included in the shipment, allow the connection of the thermometer to our laptop and the continuous data transfer.

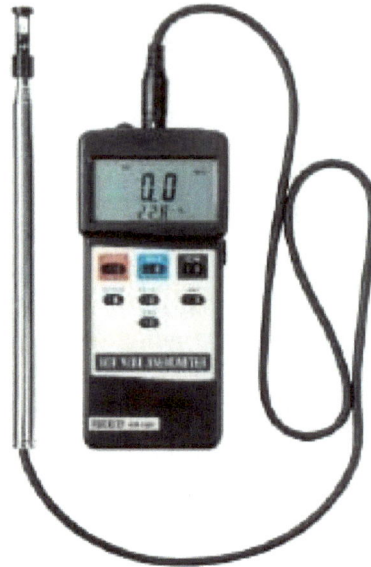

Fig. (5.6). Anemometer.

Fig. (5.7). Digital thermometers UT325.

The characteristics of the digital thermometers are given in Tables **5.1** and **5.2**:

Table 5.1. General characteristics of the digital thermometer.

Power	USB (Auto Double White Display Backlight) or 9V Battery (6F22)
LCD Size	30 × 32 mm
Product Net Weight	270 g
Product Size	175 × 85 × 30 mm
Standard Accessories	Battery, English Manual, USB Cable, Software, K Type Temperature Probe

Table 5.2. Main specifications of digital thermometer.

Basic Functions	UT325
Temperature -200.0°C to +1372°C (K-Type) -328.0°F to +2501°F	√
Temperature -210.0°C to +1200°C (J-Type) -346.0°F to +2192°F	√
Temperature Best Accuracy	K, J, T, E type: 0.2% ± 0.6°C R, S type: 0.2% ± 2°C N type: 0.2% ± 1.5°C
Display Resolution	0.1°C(When > 999.9°C : 1°C)
T1, T2 input	√
High/Low Signal Output	√
User Self-Calibration	√
Data Logging	100

3.3. Measuring Location

In the last chapter, the numerical results show that the maximum velocity locates at the chimney input. This way, we choose a possible point near the chimney inlet where we fixed the anemometer sensor. Besides, we measuring the velocity in different locations at the collector (Fig. **5.9**). We know that the collector is the main responsible for heating of air, so we fixed many sensors along the collector centerline to control the temperature. The thermocouple of type J was used to measure air temperature. We choose six typical points (L1-L2-L3-L4-L5-L6) in the collector, the distance between two successive is 0.3 m in the east section through the middle axis, that the heights of sensors can be freely adjusted, the last position is to measure the maximum velocity located at the chimney inlet (L0), the different position is shown in Fig. (**5.8**) and Fig. (**5.9**). The method of velocity measuring is shown in Fig. (**5.10**).

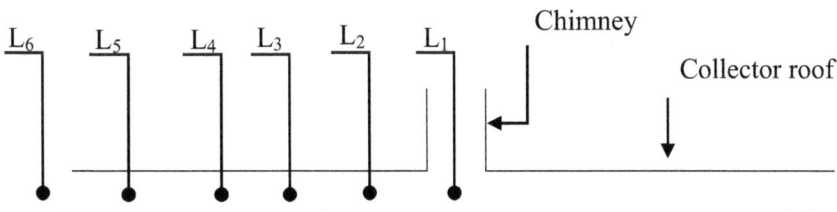

Fig. (5.8). Positions of sensors.

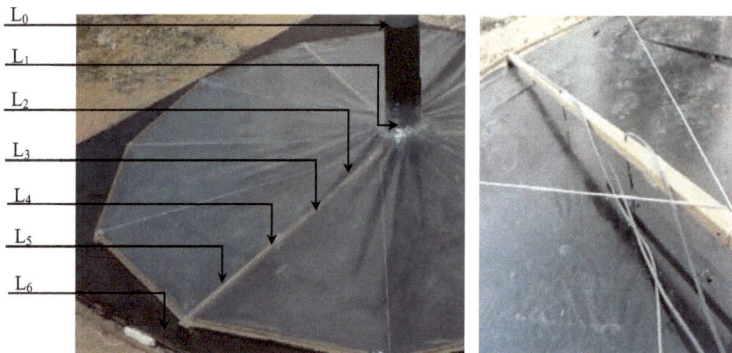

Fig. (5.9). Locations of measuring.

The coordinates of locations measuring are given in **5.3**.

Table 5.3. Points coordinates.

Point	Angle θ°	Radius r (cm)	Height z (cm)
L_0	0	0	50
L_1	0	0	0
L_2	0	30	0
L_3	0	60	0
L_4	0	90	0
L_5	0	120	0
L_6	0	150	0

(a) Collector velocity.

(b) Chimney velocity.

Fig. (5.10). Method of the velocity measuring.

3.4. Electric Power Measuring

The universal multimeter (Fig. **5.11**) is called Universal as it is used to group multiple devices in one, such a controller uses voltmeter to measure voltages DC and AC, ammeter to measure current, Ohm meter to measure resistance, plus some can monitor diodes or transistors. However, we recall the employees' baths.

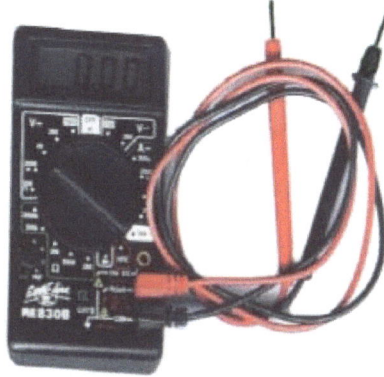

Fig. (5.11). Universal multimeter.

4. EXPERIMENTAL RESULTS

The experimental study is conducted during May and June 2016. We study the evolution of the temperature and air velocity in our system. Furthermore, the measure of the electric power produced by the turbine is responsible for estimating the performance system.

4.1. Radiation

Solar radiation is the main parameter in our experimental study of the solar chimney and it is essential to calculate the power output. To measure the solar radiation values and to record these values, we took these data from the Metrologic data of Tunisia, Sfax.

4.1.1. Daily Radiation

Fig. (**5.12**) shows the hourly variation of the global radiation for different days of May and June. According to these results, it is clear that the radiation presents the same evolution during the day for different cases. The global radiation reaches the maximum value at 12:00 h and it is minimal at the beginning and the end of the day. Indeed, the global radiation on May 23, May 28, and June 02 are slightly superior to the other days. For this reason, we propose to choose May 23, 2016, as a typical day for further study.

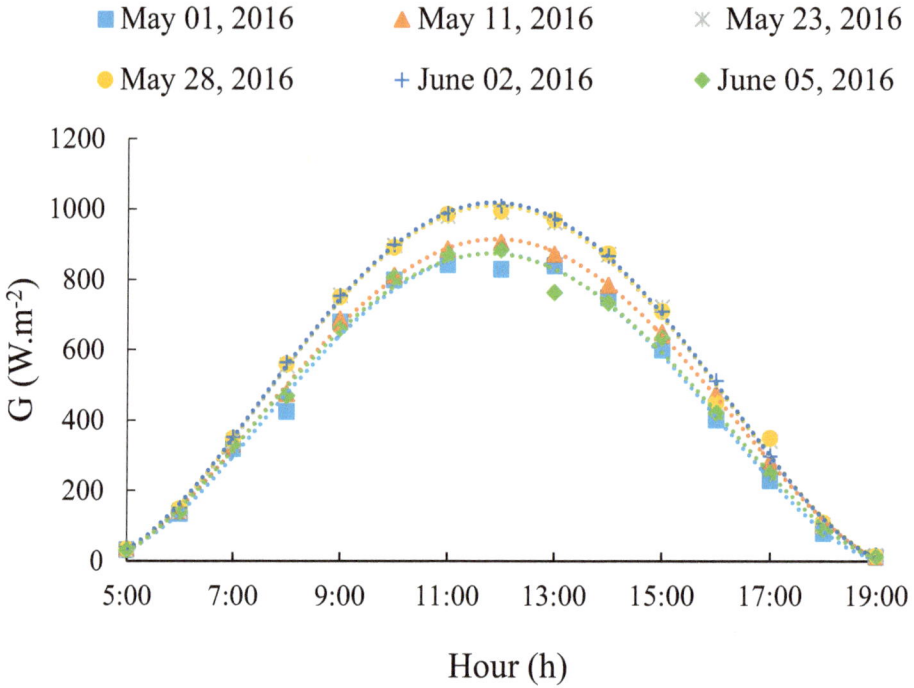

Fig. (5.12). Solar radiation for different days.

4.1.2. Typical Day

The environmental parameters such as global, diffuse, direct solar radiation, and ambient air temperature are monitored for a typical day of May 23, 2016. Indeed, the experimental data on May 23, 2016, is given by appendix 3. Fig. (**5.13**) reveals the relation between global, diffuse, direct solar radiation, and ambient air temperature of the typical day of May 23, 2016. The maximum global radiation is observed by 991 W.m^{-2} at 12:00 h, whereas, the maximum diffuse solar radiation is 146 W.m^{-2}. As far as the ambient air temperature is concerned, it found to be a maximum of 43 °C at 14:00 h and minimum by 25 °C at 8:00 h.

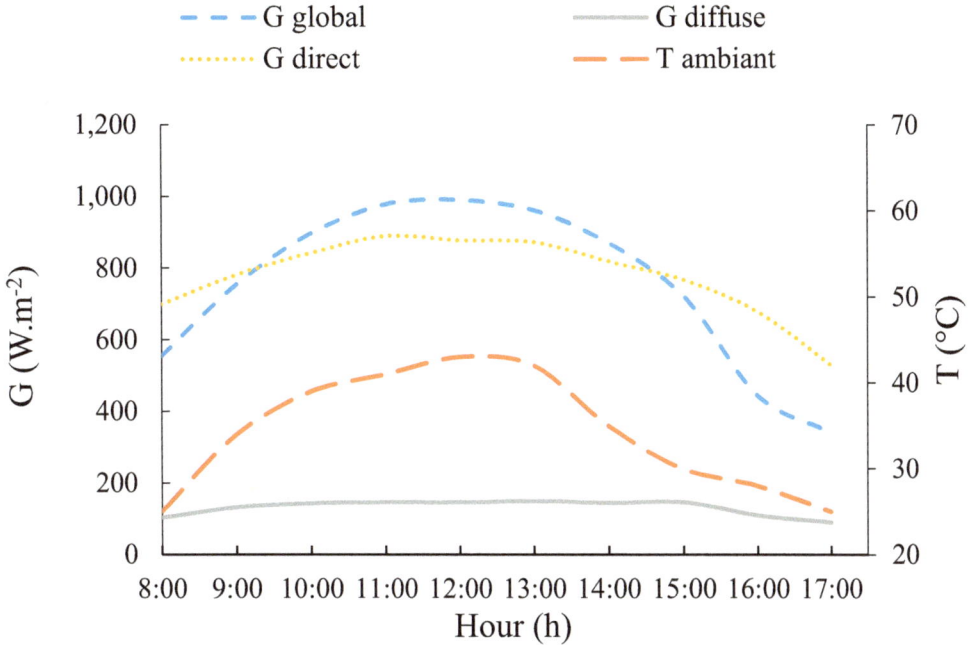

Fig. (5.13). Solar radiation and temperatures for a typical day of May 23, 2016.

4.2. Temperature

The temperature difference value between collector input and output is responsible for the airflow in the system because the heating air reduces its density and by the Archimedes floatability phenomenon the hot air rises. For the first time, we measure the daily temperature value in different points of the collector centerline in some days of May and June 2016. The second time, we study the temperature in the same position on a typical day, which we choose.

4.2.1. Daily Temperature

In this section, the experimental data of temperature is recorded on May 01, May 05, May 11, May 17, May 23, May 28, June 02, and June 05, 2016. Fig. (**5.14**) shows the temperature variation in the location L1 consisting of the chimney inlet on the day from 08:00 h to 16:00 h. According to these results, it is clear that the temperature in the chimney inlet reaches its maximum value on May 23, 2016. Besides, from these results, the temperature value is maximal at 12:00 h, due to the solar radiation is still maximal at this time. Fig. (**5.15**) illustrates the daily variation of the temperature for different positions in the collector radius at 12:00

h. According to these results, the temperature followed the same pace for all the positions and was maximum on May 23, 2016. Indeed, the air temperature increases in the direction of flow to reach its maximum at the chimney inlet.

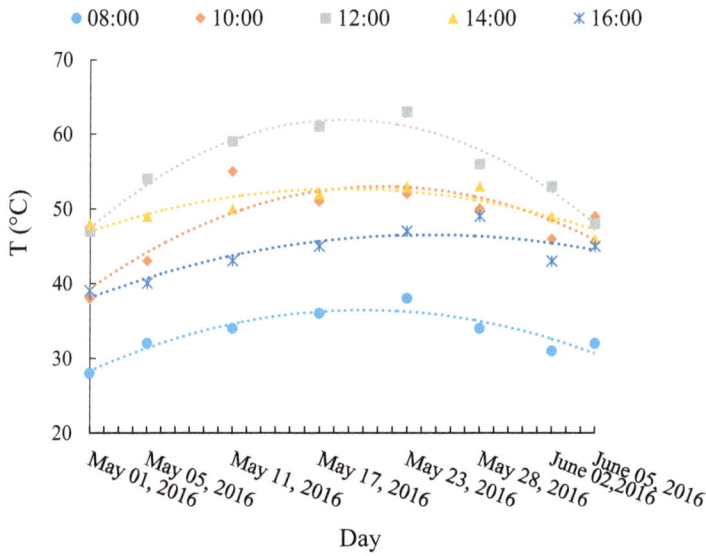

Fig. (5.14). Daily variations of the temperature in the location L1.

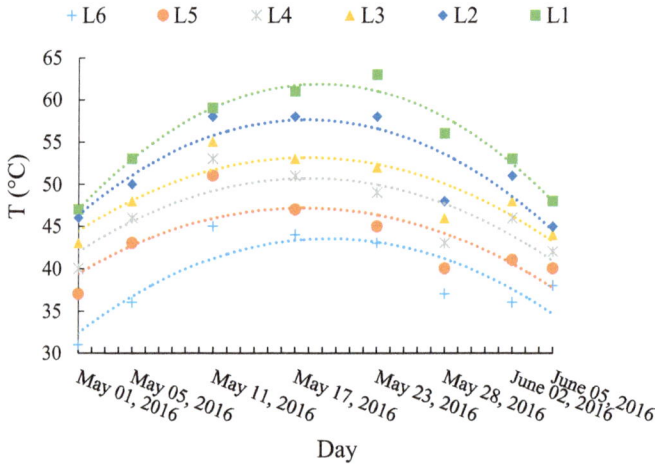

Fig. (5.15). Daily variations of the temperatures along the collector.

Fig. (**5.16**) illustrates the variation of the collector temperature for May 01, May 23, and June 02, 2016, at 12:00 h. According to these results, it is clear that the temperature presents the same distribution for all days and it is maximum on May 23, 2016. Besides, the temperature increases along the collector radius due to the greenhouse effect, as the radiation is minimal for May 01, 2016, and directly influences the temperature along with the collector, consequently to the system performance on this date. Fig. (**5.17**) shows the hourly variation of the collector temperatures for May 01, May 23, and June 02, 2016. According to these results, it is clear that the hourly temperature presents a parabolic allure since it is minimal at the beginning and the end of the day. In these conditions, the radiation is maximum at 12:00 h.

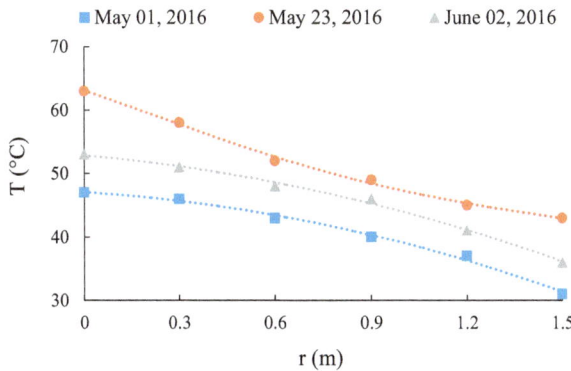

Fig. (5.16). Variation of the temperature along the collector.

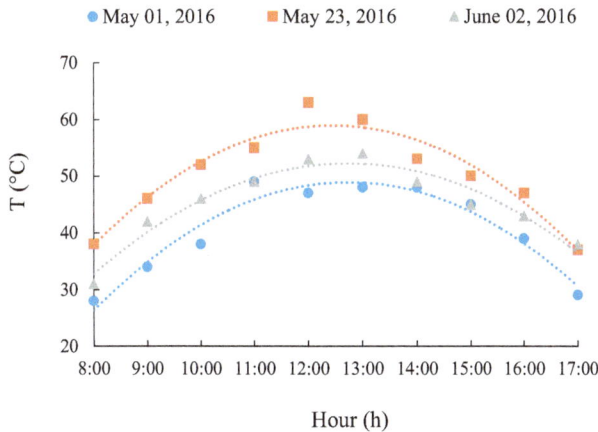

Fig. (5.17). Hourly variation of the temperatures along the collector.

The temperature difference between the entrance and the outlet collector is responsible for the airflow in the system by the buoyancy phenomenon. The hourly variation for this difference is shown in Fig. (**5.18**) for three days May 01, May 23, and June 02, 2016. According to these results, the temperature difference evolved in the same way for the three days and it is maximum at 12:00 h. The greenhouse effect is maximum at this time and it is minimal at the beginning and the end.

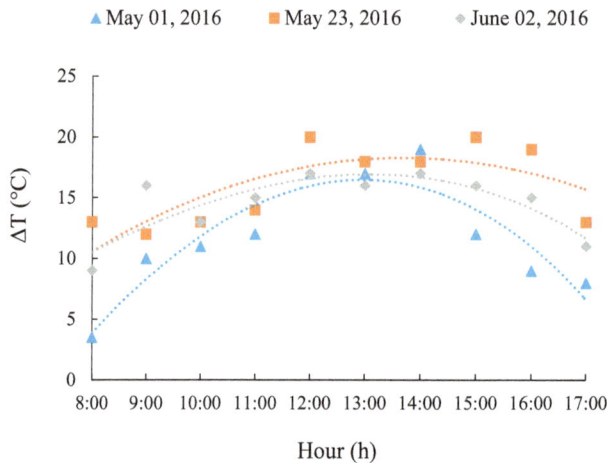

Fig. (5.18). Hourly variation of the temperature difference in the collector.

4.2.2. Typical Day

We choose May 23, 2016, as a typical day because it presents the maximum radiation when the climatic conditions are the best. Fig. (**5.19**) shows the hourly variations of the collector temperatures on May 23, 2016. According to these results, it is clear that the temperature presents the same distribution for all collector radius position. The maximum value is obtained near the chimney base (location L1) and is equal to T= 61 °C at 12:00 h since the radiation is still maximum. Indeed, it has been observed that the hourly temperature presents the same shape as the radiation graph. Fig. (**5.20**) illustrates the variation of the collector temperature for some hours. According to these results, it is clear that the temperature increases along with the collector due to the greenhouse effect and for each position that the minimum temperature at 08:00 h and the maximum temperature is reached at 12 h.

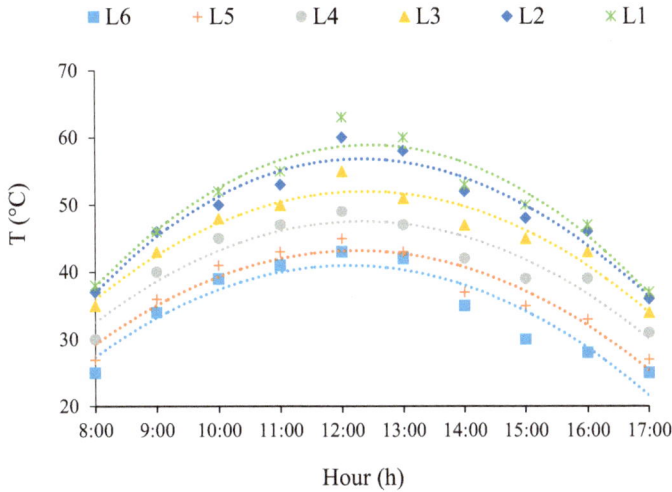

Fig. (5.19). Hourly variation of the temperatures along the collector on May 23, 2016.

4.3. Velocity

In this section, we can have interested in the measure of the daily velocity value in different points of the collector centerline in some days of May and June 2016. Indeed, we study the velocity in the same position on a typical day, which we choose.

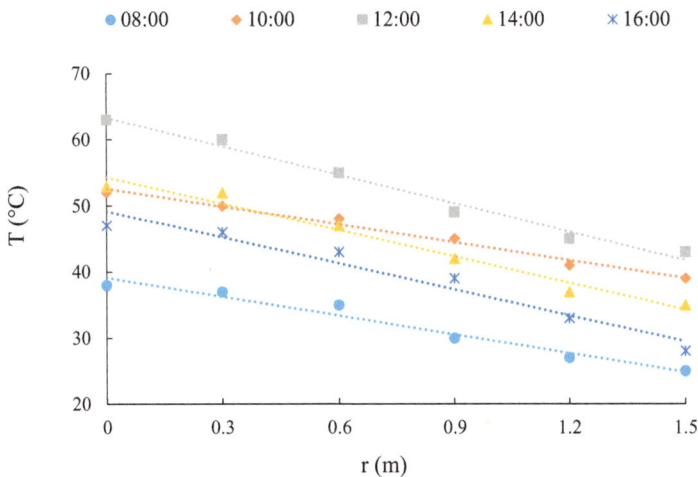

Fig. (5.20). Variation of the temperature along the collector on May 23, 2016.

4.3.1. Daily Velocity

The air velocity of the system is the most important parameter for estimating performance. Fig. (**5.21**) shows the daily variation of the collector velocity in the location L1 located in the chimney inlet. According to these results, the velocity is proportional to the temperature difference in the collector since, the temperature difference is maximum on May 23, 2016, it is evident that the velocity in the collector output is maximum on this day. As the radiation on May 01, 2016, is low, it is clear that the velocity for this day is minimal. Besides, Fig. (**5.22**) illustrates the daily variation of the collector velocity for all days. According to these results, it is shown that the velocity presents the same distribution for all collector radius positions and it is maximum on May 23, 2016.

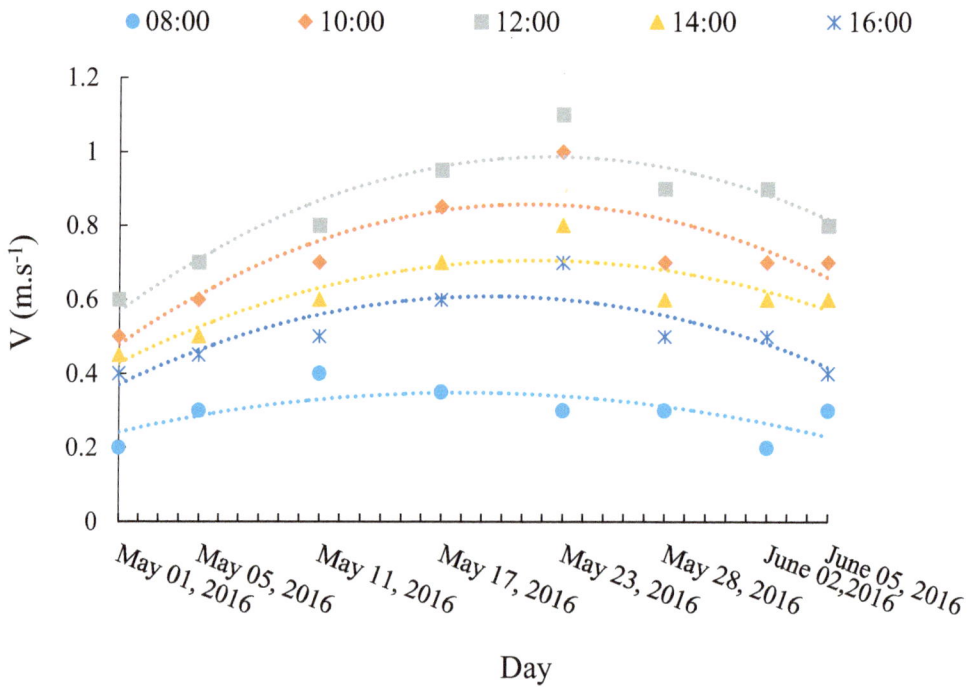

Fig. (5.21). Daily variation of the velocity in the location L1.

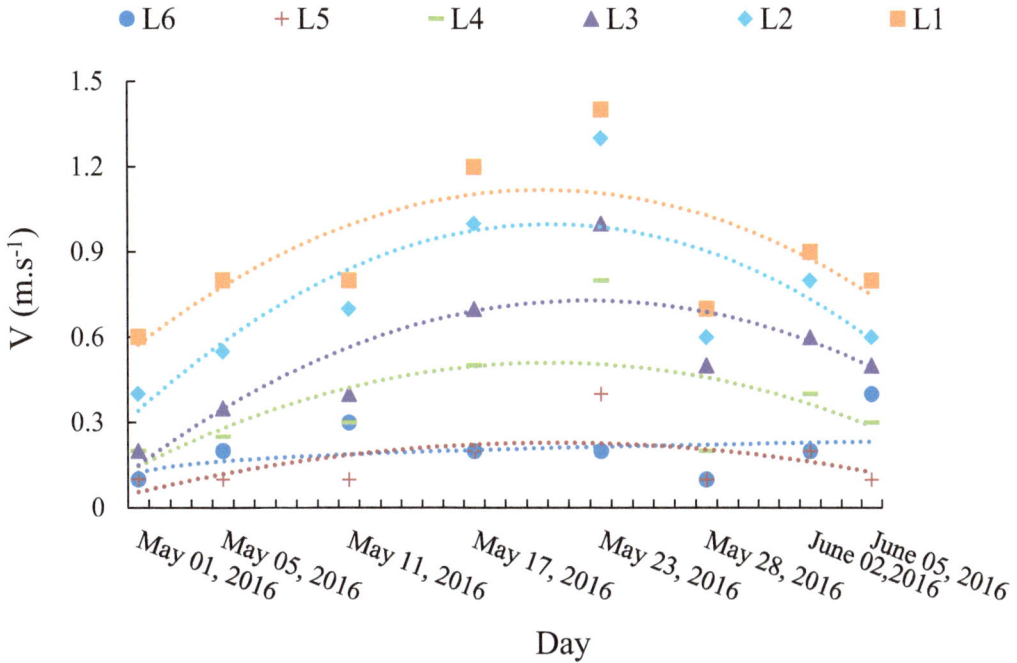

Fig. (5.22). Daily variation of the velocity along with the collector.

Fig. (**5.23**) illustrates the variation of the collector velocity on May 01, May 23, and June 02, 2016, at 12:00 h. According to these results, it is clear that the velocity presents the same distribution for all days, which the maximum is obtained on 23 May 2016. Besides, the velocity increases along the collector radius since the temperature increases and the velocity is minimal at the collector inlet since the air is not yet heated. Indeed, as the radiation is minimal on May 01, 2016, the velocity along the collector is directly influenced consequently to the system performance on this date. Fig. (**5.24**) shows the hourly variation of the collector velocity on May 01, May 23, and June 02, 2016. According to these results, it is clear that the velocity variation evolved in a parabolic shape. It is minimal at the beginning and the end of the day because the radiation is low. It is maximum at 12:00 h and reaches V=1,4 m.s^{-1} on May 23, 2016.

▲ May 01, 2016   ■ May 23, 2016   ◆ June 02, 2016

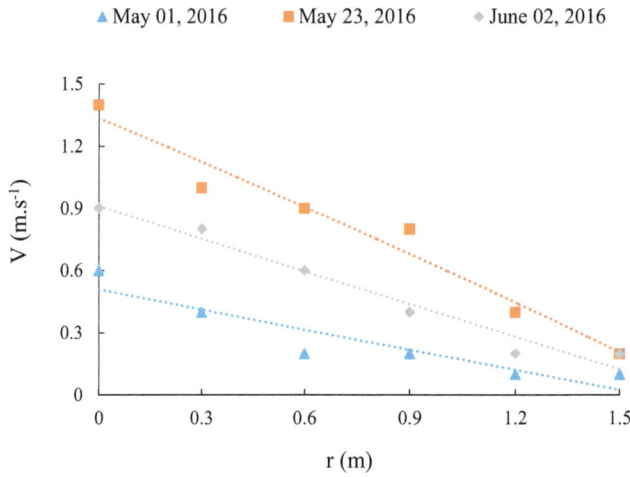

Fig. (5.23). Variation of the velocity along with the collector.

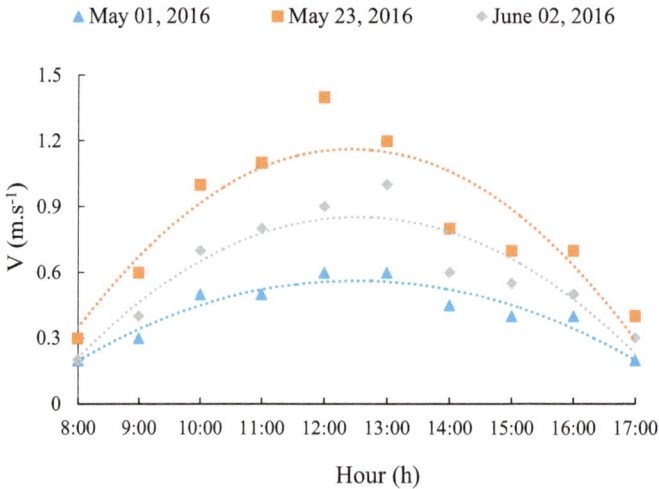

▲ May 01, 2016   ■ May 23, 2016   ◆ June 02, 2016

Fig. (5.24). Hourly variation of the velocity along with the collector.

4.3.2. Typical Day

Fig. (5.25) shows the hourly variations of the collector velocity on May 23, 2016. According to these results, it is clear that the velocity presents the same distribution for all collector radius positions the maximum value is obtained at 12:00 h since the temperature difference is maximum. Indeed, it has been observed that the hourly velocity presents the same shape of the graph for the temperature. Fig. (**5.26**) illustrates the variation of the collector velocity for some hours. According to these results, for all hours the velocity increases along with the collector due to the greenhouse effect. Besides, the minimum velocity is obtained at 08:00 h and the 17:00 h but the maximum velocity was reached at 12 hours.

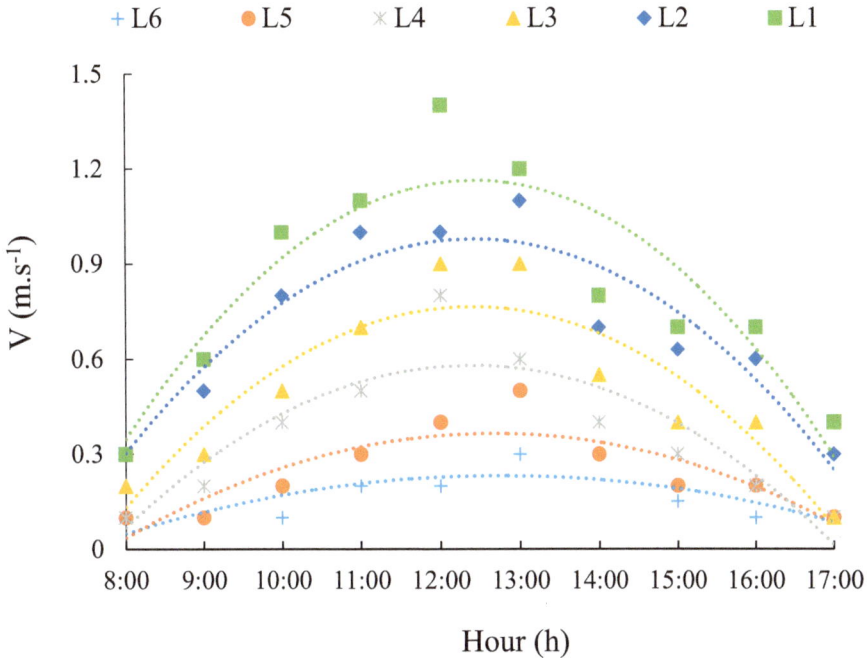

Fig. (5.25). Hourly variation of the velocity along with the collector on May 23, 2016.

4.4. Power Output

The main objective of our study is the testing of the energetic production of our prototype thus, it has been testing the power output generated by the turbine generator (Pe) on a typical day May 23, 2016. Indeed, the power has been calculated by two different methods, the power calculates with the temperature evolution along the day and the electric power measuring directly. The multiplication of the current and the voltage in the expression obtain the electric power measured (equation 1-25), the calculated power is obtained by the equation (1-24).

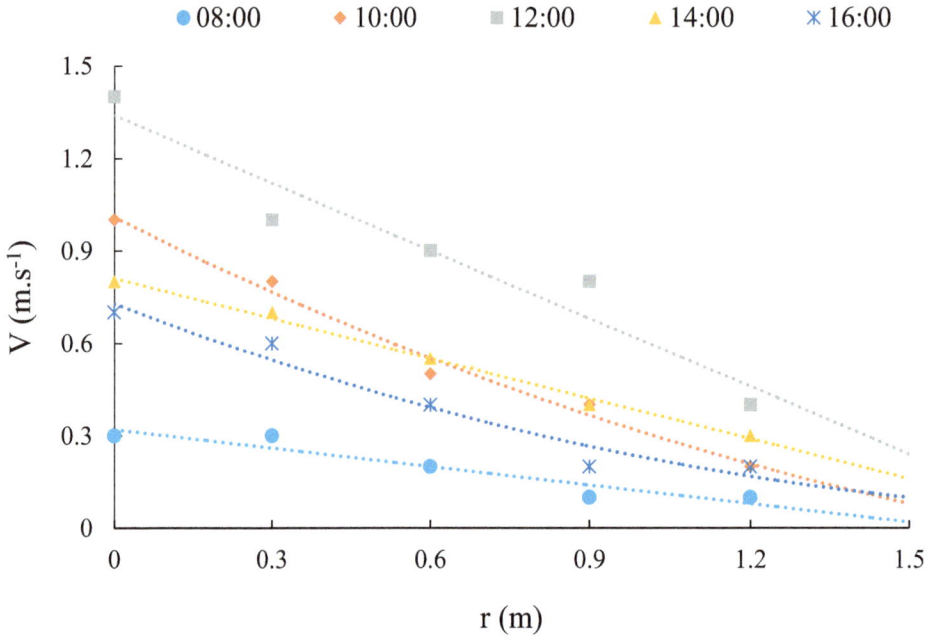

Fig. (5.26). Variation of the temperature along with the collector on May 23, 2016.

The parameters values are given in Table **5.4**.

Table 5.4. Parameters values.

	symbol	Value
Height chimney	H	3 m
Collector area	A_c	5.74 m^2
Specific heat capacity of the air	C_p	1006 J.kg^{-1}.K^{-1}
Gravity acceleration	g	9.81 m.s^{-2}
Heat transfer coefficient	h	8 W.m^{-2}
Absorptivity of the absorber surface	α	0.7
Transmissivity of the cover collector	τ	0.88
The efficiency of the turbine generator	η_{tg}	0.6

Fig. (**5.27**) shows the hourly variation of output power. According to these results, it is clear that the power reaches its maximum at 12:00 h since the radiation is maximum at this time but if the radiation decreases in the beginning at 08:00 h and the end at 17:00 h of the day, the output power decreases. Indeed, it has been observed that the profile of the calculated power is similar to the tendency of the measuring points along the day. So, it can be a result that the measuring power is dependent directly on the radiation and the difference in the temperature in the collector.

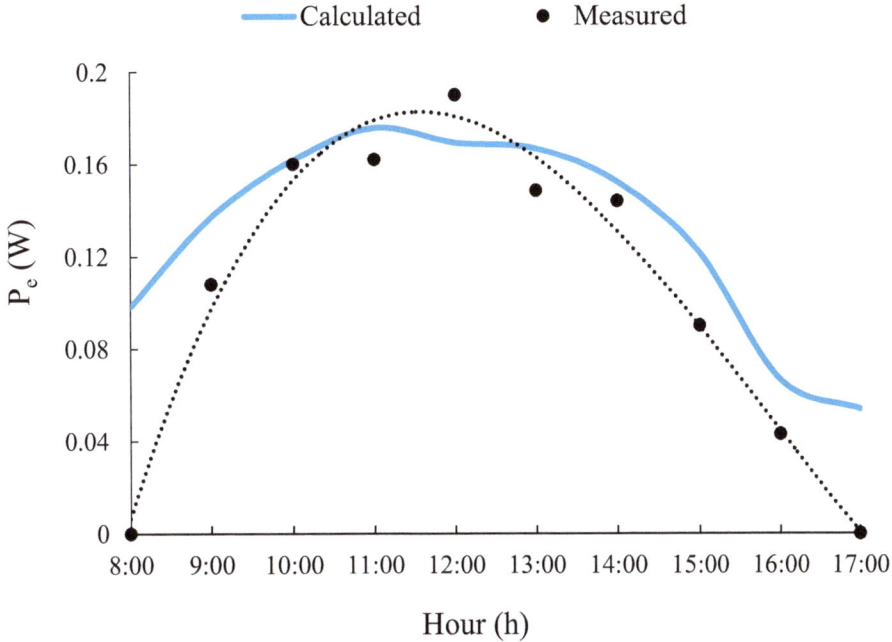

Fig. (5.27). Hourly variations of electrical power.

4.5. Correlation

In this section, we are interested in developing different experimental results to estimate some correlation of the maximum velocity and the power output of solar chimneys on the typical day of May 23, 2016, as well as to examine the effect of various conditions on the performance of the system.

4.5.1. Velocity

Fig. (5.28) shows the influence of the different temperatures on the velocity in the location L1 for a typical day of May 23, 2019, at 12:00 h. The velocity in the chimney inlet increases with the increase in the different temperatures. Indeed, it has been noted that the equation of the correlation result is similar to the theoretical equation (1-16). When the maximum velocity is proportional to the square root of ΔT. However,

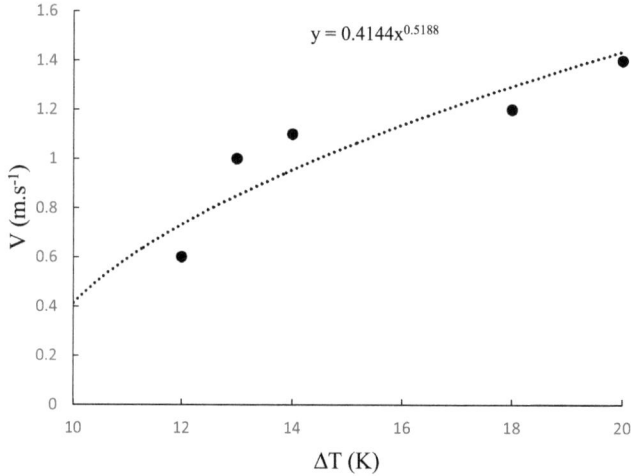

Fig. (5.28). Influence of the different temperatures on the velocity in the location L1.

we obtain by the numerical study that the maximum velocity is located near the chimney base consisting of the collector outlet. Therefore, the velocity has the maximum value in the location L1. Fig. (**5.29**) illustrates the influence of the radiation on the velocity in the location L1 for a typical day of May 23, 2019, at 12:00 h. According to these results, it has been observed that the variation of the velocity is very low for the radiation until $G=700$ W.m^{-2} and it increases for the other values. The correlations obtained show that the solar radiation and the difference of the temperature in the collector are the main parameters responsible for the airflow in the solar chimney system.

In these conditions, we have obtained this correlation:

$$V = 0,4144 \, \Delta T^{0,5188} \tag{5-1}$$

Where V is the velocity measured in the collector center.

$$V = 0,0021 \, \Delta T^{3} - 0,0495 \, \Delta T^{2} + 0,3694 \, \Delta T - 0,1455 \tag{5-2}$$

Where ΔT is the difference temperature.

$$y = 0.0021x^3 - 0.0495x^2 + 0.3694x - 0.1455$$

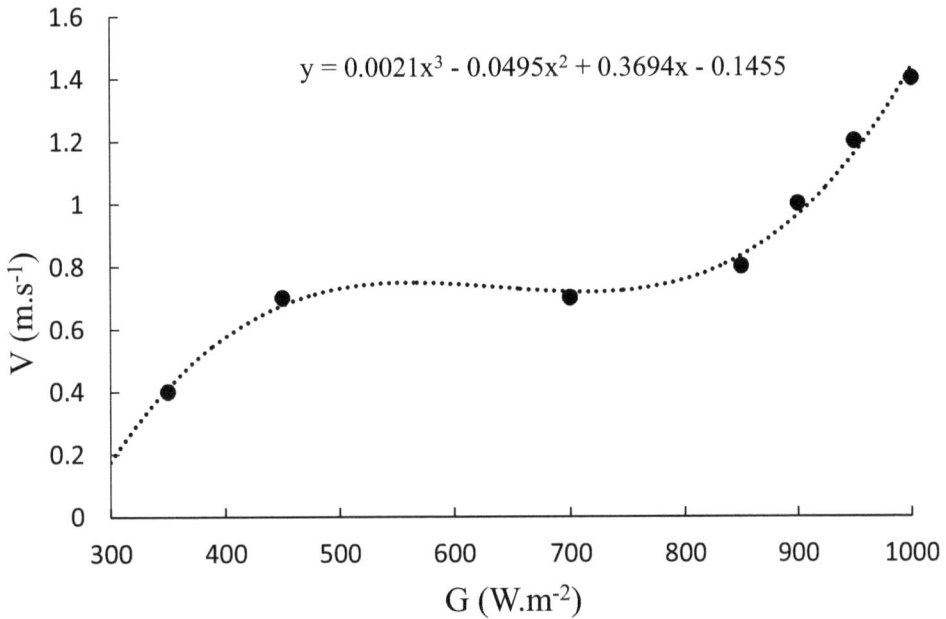

Fig. (5.29). Influence of the radiation on the velocity in the location L1.

4.5.2. Power Output

The maximum velocity in the base of the chimney where the turbine is located in the system plays a very important role in the power output of the solar chimney. Fig. (**5.30**) illustrates the influence of the maximum velocity in the location L1 on the power output for a typical day of May 23, 2019, at 12:00 h. According to these results, it has been noted that the power output increases with the maximum velocity. Indeed, the correlation given by these results shows that the power output is proportional to the square of the velocity because the eclectic power is directly proportional to the kinetic energy generated by the system. Besides, from a comparison with Fig. (**5.28**), it has been noted that the power output increases with the difference of temperature in the collector. Fig. (**5.31**) shows the influence of solar radiation on the power output for a typical day of May 23, 2019, at 12:00 h. According to these results, it has been noted that solar radiation influence directly the power output where the correlation obtained by these results is a linear equation. Indeed, it is clear that the solar radiation and the difference of temperature in the collector are the main important parameters affecting the power output. The increase in these two parameters increases the power output.

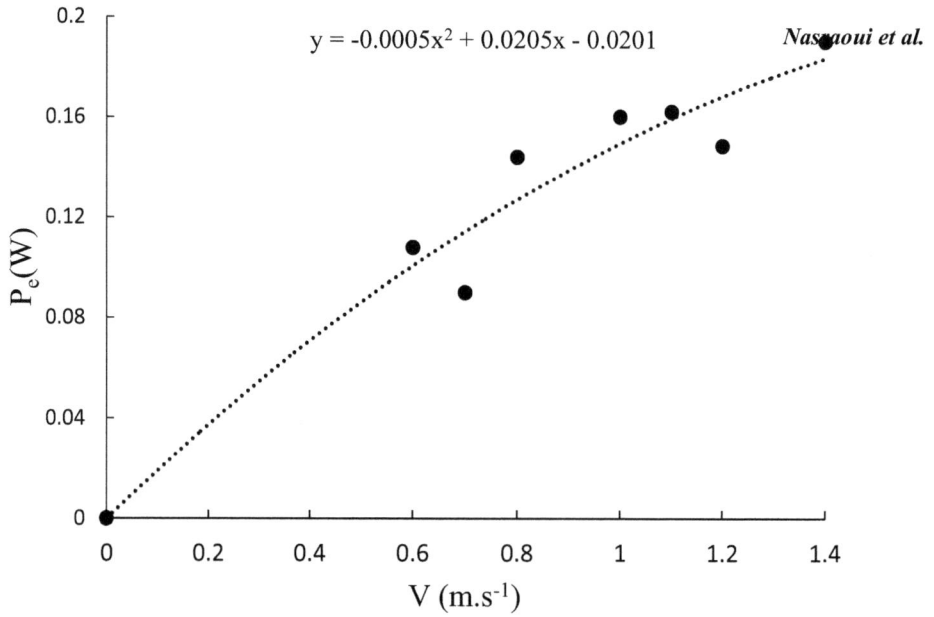

Fig. (5.30). Influence of the maximum velocity on the power output.

In these conditions, we have obtained this correlation:

$$P_e = -0,0005\ V^2 + 0,0205\ V - 0,0201 \qquad (5\text{-}3)$$

Where V is the velocity measured in the collector center.

$$P_e = 0,0131\ G - 0,0183 \qquad (5\text{-}4)$$

Where G is the measured radiation.

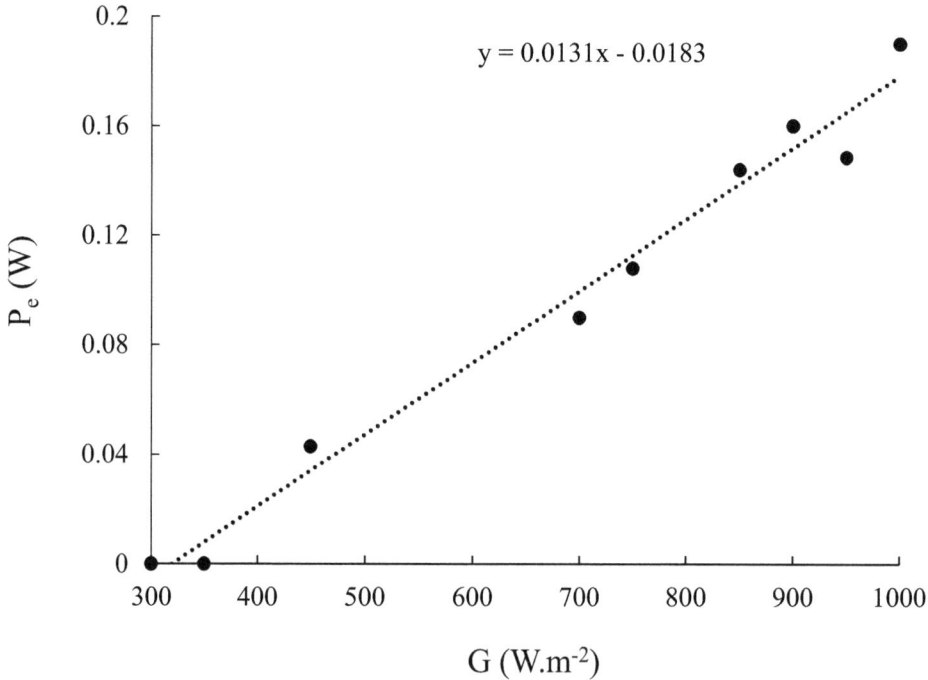

Fig. (5.31). Influence of the radiation on the power output.

CONCLUSION

The experimental study on a novel small-scale model of the solar chimney is conducted in the present chapter. Both experimental results showed that the maximum power was generated around 12:00 h for a typical day of measuring when the velocity of the rising air was greatest and as the air temperature was highest at this time. Indeed, it has been observed that the generated power has a direct relationship with the air rising velocity in the chimney base and the solar radiation when the velocity has a direct relationship with the difference of temperature in the collector. The most important result found from the present work gives a maximum velocity of 1.4 m.s^{-1}, which is remarkable, considering the small size of the prototype. However, generated power in the presented prototype is not very high because of the small scale. It is predicted that generated power will considerably increase using a larger size of this model of solar chimney.

CONCLUSION AND RECOMMENDATIONS

Solar chimney is an interesting alternative to centralized electricity generation power plants. It is very important for the future because our resources are limited, except sun and the study of this technology has a topicality in the energetic fields. In these objectives, we are interested in this book on the numerical and experimental study of a solar chimney.

At first, a detailed literature survey of this system was performed. Particularly, we have developed the computational fluid dynamics (CFD) simulation that used to model the airflow through a solar chimney. Mathematical derivations were provided to predict the theoretical equations used in CFD software to model the solar chimney.

In this part, we have concluded that the best or realistic models used to model the airflow through a solar chimney are the realizable k-ε turbulence model, and the DO radiation model and convection heat flux transfer model were employed in ANSYS Fluent. This model has been validated with anterior experimental results due to the acceptable coherence between its numerical results and experimental results. This study shows that CFD is a useful tool for investigating the possibilities of a proposed device before prototype and testing. It is helpful in the development of solar chimney retrofit for existing buildings.

In the second part, alternate geometric configurations of solar chimneys were studied numerically to expand the study of optimizing the design of the solar chimney. Simulations were performed using the commercial CFD software ANSYS fluent. The goal of this part is the study of the effect of the geometric parameters on the airflow behavior inside the solar chimney to obtain an optimal size available to construct a prototype of solar chimney. This study shows that the increase in the height, in the diameter of the chimney, and the diameter of the collector increases the temperature and the air velocity however, an increase in the collector height decreases.

An experimental study is presented in the last part of this book. The experimental prototype, constructed at ENIS, is used to study the environmental temperature, distribution of the temperature, air velocity, and the power output generated by the turbine. The main results were found from this prototype are the solar radiation and the gap of temperature in the collector. These parameters are important factors affecting the performance of the solar chimney. It is found that the temperature gap in the collector is approximately equal to 20 °C when the solar radiation is great at noon. It is therefore concluded that the solar chimney stands to achieve good performance in areas that exhibit the highest monthly average daily solar radiation. Indeed, we have concluded that our country, particularly the region of Sfax, presents a good area for installing the solar chimney power plant, where the technology and the material to build such plants are available.

Based on the conclusions, the following suggestions and recommendations might be found useful in future work:

- Increasing the height of the collector and the diameter of the chimney to improve the power output and efficiency.
- Installing a storage heat system, it has been suggested that by placing extra thermal mass under the collector in the form of black containers with water, the plant could generate power at night.
- Studying the effect of different materials on absorber and the collector to determine the best material that give the most performance.

References

Askari Mohammad, B, Mirzaei, MAV & Mohsen, M (2015) Types of Solar Cells and Application. *American Journal of Optics and Photonics,* 3, 94-113.

Austin, C, Borja, R & Phillips, J (2007) Operation Solar Eagle: A Study Examining Photovoltaic (PV) Solar Power as an Alternative for the Rebuilding of the Iraqi Electrical Power Generation Infrastructure, Naval Postgraduate School. *June 2005 Last Retrieved March 30.*

Calder, R (1970) *Leonardo & the age of the eye* William Heinemann Ltd., London 106.

Chikere, AO, Al-Kayiem, HH & Abdul Karim, ZA (2011) Review on the Enhancement Techniques and Introduction of an Alternate Enhancement Technique of Solar Chimney Power Plant. *J Appl Sci (Faisalabad),* 11, 1877-84.
[http://dx.doi.org/10.3923/jas.2011.1877.1884]

Dhahri, A (2013) A Review of solar Chimney Power Generation Technology. *International Journal of Engineering and Advanced Technology (IJEAT),* 2, 1-17.

DLR (Deutsches Zentrum für Luft- und Raumfahrt - German Aeroposace Center) (2005) *Concentrating Solar Power for the Mediterranean Region, MED-CSP, Final Report, DLR, Institute of Technical Thermodynamics, Stuttgart, Germany.*

Haaf, W (1984) Solar chimneys, Part II: preliminary test results from the Manzanares pilot plant. *Int J Solar Energy,* 2, 141-61.
[http://dx.doi.org/10.1080/01425918408909921]

Hassanien, RHE, Li, M & Lin, WD (2016) Advanced applications of solar energy in agricultural greenhouses. *Renew Sustain Energy Rev,* 54, 989-1001.
[http://dx.doi.org/10.1016/j.rser.2015.10.095]

Alibakhsh, K, Mehran, G & Mehrdad, G (2014) Simulation and optimization of geometric parameters of a solar chimney in Tehran. *Energy Conversion and Management,* 83, 28-34.

Kaltschmitt, M, Streicher, W, Wiese, A (Eds.) (2007) *Renewable energy: Technology economics and environment.* Springer-Verlag Berlin Heidelberg.

Kreetz, H (1997) *Theoretische Untersuchungen und Auslegung eines temporären Wasserspeichers für das Aufwindkraftwerk, diploma thesis, Technical University Berlin, 1997.*

Ketlogetswe, C, Fiszdon, JK & Seabe, OO (2008) Solar chimney power generation project the case for Botswana. *Renewable and Sustainable Energy Reviews,* 12(7), 2005-12.
[http://dx.doi.org/10.1016/j.rser.2007.03.009]

Koonsrisuk, A & Chitsomboon, T (2013) Effects of flow area changes on the potential of solar chimney power plants. *Energy,* 51, 400-6.
[http://dx.doi.org/10.1016/j.energy.2012.12.051]

Krisst, RJK (1983) Energy transfer system. *Alternative Sources of Energy,* 63, 8-11.

Kulunk, H (1985) A prototype solar convection chimney operated under Izmit conditions. In: Veiroglu, TN, (Ed.), *Proceedings of the 7th Miami international conference on alternative energy sources,* 162.

Messenger, R & Ventre, J (2000) *Photovoltaic Systems Engineering.* CRC Press, New York 63-4.

Pretorius, J (2004) Solar Tower Power Plant Performance. *Thesis, University of Stellenbosch.*

Pastohr, HK (2003) Numerical and analytical calculations of the temperature and flow field in the upwind

power plant. *International Journal of Energy Research,* 28, 495-510.
[http://dx.doi.org/10.1002/er.978]

Papageorgiou, C (2006) Floating Solar Chimney Technology. *Solar Energy.*
[http://dx.doi.org/10.5772/8069]

Penick, T & Louk, B (2007) Photovoltaic Power Generation, TEI Controls. *December 1998 Last Retrieved March 30.*

Schlaich, J (1995) *The Solar Chimney: Electricity from the Sun.*Edition Axel Menges, Stuttgart, Germany.

Sherif, SA, Pasumarthi, N, Harker, RA & Brinen, GH (1995) Performance of a demonstration solar chimney model for power generation. *Final Technical Report No UFME/SEECL-9507, Solar Energy and Energy Conversion Laboratory, Department of Mechanical Engineering, University of Florida, Gainesville, Florida.*

Zhou, XP, Wang, F & Fan, J (2010) Performance of solar chimney power plant in Qinghai–Tibet Plateau. *Renewable and Sustainable Energy Reviews,* 14, 2249-55.
[http://dx.doi.org/10.1016/j.rser.2010.04.017]

Zhou, XP & Yang, JK (2008) Temperature field of solar collector and application potential of solar chimney power systems in China. *Journal of the Energy Institute,* 81(1), 25-30.
[http://dx.doi.org/10.1179/174602208X269364]

SUBJECT INDEX

A

Absorber 11, 31, 32, 57, 58, 59, 74, 77, 78, 79,
  80, 81, 95, 98, 104, 166
  bed 31
  interface 11
Absorption coefficient 21, 55, 104
  effective 21
  spectral 55
Air 12, 14, 15, 17, 18, 20, 21, 22, 30, 31, 39,
  58, 67, 69, 79, 93, 104, 133, 164, 166,
  173, 187
  ambient 14
  cold 39
  collector 31
  energy 166
  friction 93
  heating 14, 173
  movement 69
  pressure 58
Air heat 74, 122
  transfer 122
ANSYS 34, 39, 40, 53, 55, 57, 58, 74, 97, 188
  design modeler 40
  fluent 34, 39, 55, 57, 58, 74, 97, 188
  meshing 40
Application 6, 7, 11, 16, 34, 40, 42, 57, 167
  energy storage 6
Archimedes floatability phenomenon 173
Atmosphere 13, 22, 25
  surrounding 22
Atmospheric pressure 98, 106, 116, 126, 136,
  145, 146, 152
  static 98
  value 116, 136, 145

B

Bernoulli's theorem 152
Bizarre contraption 25
Black body 55
  intensity 55

Botswana's ministry of science and
  technology 31
Boundary conditions 42, 56, 58, 59, 74, 83,
  97, 98
  and numerical parameters 97
  parameters 98
Boundary layer 52, 54, 81
  development 52
  thermal 81
Boussinesq 33, 37, 45
  approximation 23
  model 37
  relationship 45
Buoyancy 25, 35, 37, 39, 44, 49
  force 35, 37
  -induced flow 39

C

Calculation 43, 99
  iterative 43
Cells 5, 6, 11, 42, 43, 61, 94, 97
  elementary mesh 42
  photovoltaic 5
CFD 39, 40, 41, 42, 59, 60, 63, 97
  parameters 59
  setup application 40
  simulation 41, 60, 63, 97
  solvers 42
  strategy 39
CFD code 34, 56, 57, 95
  ANSYS fluent 34
  validated 95
CFD software 39, 95, 188
  ANSYS fluent 95
Chimney 12, 14, 15, 16, 17, 20, 23, 25, 30, 57,
  59, 65, 67, 68, 69, 72, 73, 78, 79, 86, 88,
  90, 95, 98, 104, 106, 107, 109, 112, 116,
  120, 22, 126, 127, 130, 136, 138, 140,
  145, 148, 150, 162, 164, 169, 170, 173
  axisymmetric solar 98
  convergent-top 16
  divergent-top 16
  diameter effect 130

concentrator solar 2
Power stations 7, 8, 16
    hydroelectric 16
    parabolic trough 8
    thermodynamic solar 7
Pressure 10, 12, 15, 22, 23, 24, 34, 39, 59, 67,
        69, 78, 90, 98, 126
    external 78
    inlet 59, 98
    loss 22
    outlet 59, 98
    outlet boundary condition 59
Problems 11, 19, 39, 57, 59, 62, 63, 159
    constructive 159
    engineering 39
    logistical 19
    resolving complex fluid flow 39
Profiles 27, 30, 161
    numerical 161
    solar insolation 30
    turbine blade 27
Projects 28, 39, 167
    electronic 167
    enhanced process 39
    largest 28
Prototype 26, 30, 95, 96, 97, 99, 101, 103,
        159, 161, 162, 163, 165, 166, 187, 188
    pilot 26
    plant 30

R

Radiation 2, 14, 20, 54, 55, 70, 71, 75, 77, 78,
        79, 83, 87, 88, 99, 164, 171, 172, 175,
        179, 182, 184
    diffuse 99
    direct 99
    global 20, 171, 172
    infrared 164
    long-wave 14
    non-gray 55
    model theory 54
Radiative transfer equation (RTE) 54, 55
Rayleigh 38, 39
    measures 38
    number events 39
Rays 2, 8, 9, 13
    cosmic 2
    solar 8, 9, 13
Renormalization group theory 48

Reynolds number 38, 49, 53, 54
    critical 54
    effective 49
    momentum-thickness 53
    transition onset momentum thickness 53
Reynolds stress, normal 45
Reynolds stresses 45
RNG 48, 49, 50, 51
    model 48, 50, 51
    theory 48, 49, 50, 51

S

Simulation(s) 34, 39, 57, 58, 59, 60, 61, 63,
        74, 83, 99, 159, 188
    computer 74
    converged 59
    numerical 39, 57, 58, 61, 159
    parameters 99
    software 39
    steady-state 59
Soil 17, 18, 30, 32
    compacted 32
    natural 30
    sprayed 17
    wet 17
Solar 1, 2, 3, 6, 7, 9, 25, 33
        aero-electric power plant in north Africa
        25
    resource 1
    spectrum 2, 3
    supplement 7
    systems 3, 5
    technology 33
    tower power plants 9
    water-heaters 6
Solar chimney 1, 7, 11, 12, 14, 17, 18, 19, 20,
        25, 28, 29, 33, 34, 57, 62, 71, 72, 74, 77,
        79, 82, 84, 92, 95, 114, 129, 131, 133,
        139, 142, 150, 152, 157, 159, 162, 187,
        188, 189
    collector 20
    design 142
    functions 17
    operation 84
    photovoltaic nuclear 19
    power plant 7, 11, 12, 19, 29, 34, 131, 159,
        189
    retrofit 188
    shape 95

www.ingramcontent.com/pod-product-compliance
Lightning Source LLC
Chambersburg PA
CBHW050846220326

41598CB00006B/451